International Comparative Social Studies

VOLUME 62

The titles published in this series are listed at *brill.com/icss*

Sufficiency

Sufficiency

From Growth and Overshoot to Enoughness

Edited by

Toni Ruuska
Tina Nyfors

BRILL

LEIDEN | BOSTON

Chapter 10 of this publication, titled *Taking Population Seriously in the IPAT Equation: Ethical and Political Implications of Planetary Boundaries*, authored by Michel Bourban, is available under an Open Access CC BY license, supported by the Behavioural, Management and Social sciences (BMS) OA Fund, University of Twente, Enschede, The Netherlands, Work Order No. 14535000-10, Philosophy Section.

This publication was supported by funding from the Research Council of Finland (Project Skills of Self-Provisioning in Rural Communities, Grant No. 343277), the Kone Foundation (Project Underdogs of Just Transition, Grant No. 202302807), and the Tiina and Antti Herlin Foundation..

Cover illustration: Landscape becoming hybrid © Risto Musta.

Typeface for the Latin, Greek, and Cyrillic scripts: "Brill". See and download: brill.com/brill-typeface.

ISSN 1568-4474
ISBN 978-90-04-74639-8 (hardback)
ISBN 978-90-04-74640-4 (e-book)
DOI 10.1163/9789004746404

This book is printed on acid-free paper and produced in a sustainable manner.

Contents

PART 3
Sufficiency in Technology

PART 4
Sufficiency in Population

Conclusion

Acknowledgements

The original spark of this book came from the 4th Peaceful Coexistence Colloquium, *Sustainable Change and Sufficiency: Philosophies, practices, pathways*, which was held online in December 2021.

The editors would like to thank and acknowledge the commitment and input of all the authors who have contributed to this book. From Brill, the editors would like to thank Simona Casadio for her guidance and professionalism. In addition, the editors wish to acknowledge the support and encouragement of many of our colleagues at the University of Helsinki. Special mention is due to Professor Pasi Heikkurinen for his initial inspiration for this book and for his on-going intellectual, moral, and other support for our collective work.

Toni Ruuska wants to thank the funders of his work. His work was supported by funding from the Research Council of Finland, project "*Skills of self-provisioning in rural communities*" (SOS), grant number 343277, and Kone Foundation, project "*Underdogs of just transition*" (*Oikeudenmukaisen siirtymän altavastaajat*), grant number 202302807. The work of Tina Nyfors was supported by the Tiina and Antti Herlin Foundation.

Finally, the editors would like to acknowledge the efforts of the Sustainable Change Research Network (SUCH); as a vibrant community of like-minded scholars, SUCH has provided invaluable assistance to our research work.

Figures and Tables

Figures

Tables

Notes on Contributors

Editors:

Toni Ruuska
is University Lecturer at the Centre for Consumer Society Research, Faculty of Social Sciences, University of Helsinki; Principal Investigator of *Underdogs of Just Transition* research project (2024–2028); Co-editor of Mayfly Books. Ruuska has co-edited *Sustainability beyond Technology* (Oxford, 2021) and is the author of *Reproduction Revisited: Capitalism, Higher Education and Ecological Crisis* (Mayfly, 2019). In addition to wide-ranging critique of capitalism, and technology, his research deals with degrowth, self-provisioning, and alternatives to capitalism. Theoretically he is involved in ecological Marxism, eco-feminism, and degrowth.

Tina Nyfors
is Doctoral Researcher at the University of Helsinki. Her research focuses on sufficiency and her PhD thesis deals with sufficiency in the policy and environmental movement contexts. After the thesis is published, she will continue as a postdoctoral researcher. She has published, e.g., the article Ecological Sufficiency in Climate Policy: Towards Policies for Recomposing Consumption (Futura, 2020).

Contributors (alphabetical):

Karl-Johan Bonnedahl
PhD, Docent in Business administration at Umeå University. He is positioned within strong sustainability scholarship, and interested in conceptualizations, assumptions and ethics related to economic activities and the policy and discourse of sustainable development. Currently he's doing applied research on social innovation with a focus on human-nature interaction and dependency, and on impact management, with a focus on mission-driven entrepreneurship and counselling.

Michel Bourban
is Assistant Professor in ethics and political philosophy at the University of Twente. He works on fields such as climate justice, environmental ethics,

animal ethics, and citizenship theory and is interested in topics such as sustainability, eco-anxiety, energy transition, and human rights. His research line, "Extending Justice and Citizenship", aims at clarifying the demands of justice and citizenship in our aera of global environmental changes.

Philip Cafaro
recently retired as Professor of philosophy at Colorado State University, USA and is co-Principal Investigator for The Overpopulation Project at Göteborg, Sweden. His work centers on environmental ethics, population and consumption issues, and the preservation of wild nature. He is a life member of the Sierra Club, a former ranger with the National Park Service, and past president of the International Society for Environmental Ethics. Cafaro is the author of *Thoreau's Living Ethics: Walden and the Pursuit of Virtue*, and *How Many Is Too Many? The Progressive Argument for Reducing Immigration into the United States*.

Pasi Heikkurinen
is Professor of Sustainable Business at Lappeenranta-Lahti University of Technology; Chair of The Finnish Society for Environmental Social Science; Adjunct Professor of Sustainable Economy at the University of Helsinki; Adjunct Professor of Sustainability and Organizations at Aalto University; and Co-Founder of Sustainable Change Research Network. His sustainability research focuses on questions concerning the economy (especially degrowth), technology, and culture.

Mikko Jalas
is Associate Professor in Co-Innovation for Circular Solutions at the Department of Design at Aalto University School of Arts, Design and Architecture. His research focuses on consumption practices in affluent industrial countries, as well as time-use and societal rhythms. This line of research contributes to the critical questions of escalating demand and fair allocation of consumption opportunities. Empirically, his work has documented patterns of domestic energy use and other resource-intensive consumption practices.

Jessica Jungell-Michelsson
holds a PhD in sustainability sciences from the University of Helsinki. Her research has focused on strong sustainability and sufficiency, particularly in the food sector. Jungell-Michelsson currently works as an independent researcher and business sustainability advisor in Finland.

Lassi Linnanen
is Professor of Environmental Economics and Management in Lappeenranta–Lahti University of Technology, Lahti Campus. He has over 30 years of international experience in solving sustainability challenges in academia, in start-up businesses and in policy design. Currently he is also the Chair of Finnish Expert Panel for Sustainable Development, an independent science panel operating under the auspices of the Prime Minister's Office of Finland. His research over the years has covered a variety of topics, including sustainability transitions of energy, food and textile systems, sustainable innovations, corporate responsibility strategies, life cycle assessment, environmental and energy policy, and sufficiency as a solution for polycrisis.

J. Mohorčich
is Assistant Professor at Lehman College, where he teaches courses in American politics, political philosophy, and science policy. His work discusses transport in an energy-constrained world, the effects of constructing global networks for high-capacity computing, the capacity of artificial intelligence to reproduce poetic conventions, and whether GMOs, nuclear power, and biofuels can be used to understand the development of cell-cultured meat. He has previously taught at Johns Hopkins University and the University of Mannheim. He is from Soldotna, Alaska.

Iana Nesterova
PhD, is Visiting Researcher at Roskilde University, Denmark, and a long-term practitioner of voluntary simplicity. Her research focuses on business and lived experiences of degrowth transformations. She is the author of the book *Degrowth, depth and hope in sustainable business: Reflections from Denmark, Finland and Sweden*, and a co-author of the book *Deep transformations: A theory of degrowth.*

Thomas Princen
is Faculty Member at the School for Environment and Sustainability at the University of Michigan, Ann Arbor, Michigan, USA. His research ranges from the distancing of commerce to overconsumption, from ecological rationality to sufficiency. He currently works on the politics of urgent transition with projects on localization, fossil fuels, and extreme events. Among his books are *The Logic of Sufficiency*; *Treading Softly: Paths to Ecological Order*; *Confronting Consumption*; *Ending the Fossil Fuel Era*; *The Localization Reader: Adapting to the Coming Downshift*; *Fire and Flood: Extreme Events and Social Change, Past, Present, Future*, all published by MIT Press.

Joonas Uotinen
works as a researcher at the Finnish Institute for Health and Welfare, focusing on happiness, sustainable well-being, and social sustainability across various cultures, disciplines, and methodologies. His interests include the individual, social, institutional, cultural, and interspecies aspects of happiness and well-being, aiming to support global sustainable well-being and the equal distribution of well-being.

Tere Vadén
PhD, Associate Professor, is a philosopher and founding member of BIOS Research Unit, where he works as a researcher on the effects of ecological crises and resource limits on Finnish society. He has worked as non-tenured professor in philosophy, art education and interactive media and is a long-time editor of the Finnish philosophical journal *niin & näin*. Vadén has published several monographs, *including Heidegger, Zizek and Revolution* (2014) and *Energy and Experience* (2015, with Antti Salminen).

PART 1

Setting the Stage

CHAPTER 1

Introduction: Sufficiency in the Time of Growth and Overshoot

Toni Ruuska and Tina Nyfors

Over fifty years ago, Paul Ehrlich and John Holdren famously suggested that human-induced environmental impact (I) comes down to three factors: population (P), affluence (A), and the level of technology (T).[1] Ever since its introduction in 1972, the I = PAT equation has been surrounded by a manifold discussion,[2] and although the equation can be considered a useful heuristic tool for assessing the human-induced impacts on the environment, it is at the same time true that it is generic and reductive. Moreover, the equation was not developed to analyze the different relations of its parts or to explain causalities, priorities, or disparities.[3] For instance, the I = PAT equation does not tell us how the increasing standard of living and mounting technology are linked in the contemporary world economy, why the size of the human population keeps on rising, or what kind of cultural dynamics or historical circumstances lay behind these developments.

In any case, in the mid 2020s, we inhabit a world of severe human-induced ecological overshoot.[4] The ongoing ecological crisis has not been produced by all humans. In fact, the situation is so far from this that almost the opposite is true. Most of the global resource consumption takes place in the affluent countries and communities—the "overconsuming class"—while the rest of the world population consumes very little (see Chapter 4). The economic growth-driven world economy is one with great disparities and socio-economic inequality, a historical world system where the affluence and technologization

1 P. Ehrlich and J. Holdren, "Critique," *Bulletin of the Atomic Scientists* 28, no. 5 (1972).

2 M. Chertow, M. "The IPAT and its Variants," *Journal of Industrial Ecology* 4, no. 4: 13–29 (2001).

3 R. York, E. Rosa, and T. Dietz, "STIRPAT, IPAT and ImPACT: analytic tools for unpacking the driving forces of environmental impacts," *Ecological Economics* 46, no. 3: 351–365 (2003).

4 W. Steffen, W. Broadgate, L. Deutsch et al., "The trajectory of the Anthropocene: The great acceleration," *Anthropocene Review*, 2(1): 81–98 (2015); W. Rees, "Ecological Economics for Humanity's Plaque Phase," *Ecological Economics* 169, 106519 (2020); K. Richardson, W. Steffen, W. Lucht et al., "Earth beyond Six of Nine Planetary Boundaries," *Science Advances* 9, no. 37 (2023).

 | DOI:10.1163/9789004746404_002

of the wealthy part of humanity is secured by both the labor of others and access to resources and technology.[5]

From the basis of the existing power imbalances and systemic inequality, it can be argued that we should pay more attention to the factors of affluence (A) and technology (T), rather than paying attention to population (P), as the root causes of the ecological overshoot. However, this does not mean that we should withdraw from treating population as a sustainability challenge, but rather, it indicates that the three factors should not be treated as equals and that we should stay alert to their ethical-political underpinnings. Further caution concerning the population factor is even more justified today due the rise of the alt- and far-right, which deem population growth as the primary driver for the ecological overshoot, while promoting discriminating population reduction demands.[6]

To be clear, our intention in this book is not to push for yet another Malthusian argument. In contrast, we ask what it would mean if we applied the concept of sufficiency to affluence, technology, and population. We examine our polarized world where some have so much (too much affluence) and others so little, and solutions are proposed from the point of view of redistribution, relocalization, deep transformation, and strong sustainability. We study how industrial and fossil-fueled technology contributes to the ecological overshoot and to economic growth (too much technology), and consider possible means to alleviate the many problems of resource-intensive modern technology. After discussing affluence and technology, we then ask if it is possible to approach the question of population in a reflexive and inclusive manner. By structuring the main themes of the book from the point of view of sufficiency—and in the order of affluence, technology, and population—we, the editors, would thus like to rearrange the I = PAT equation into a new form: $I = ATP$.

To guide our inquiry in a world of too much and too little, we utilize of the humble concept of *sufficiency*, which to us entails an ethos of *enoughness* and the notion that enough is enough or, alternatively, that too much leads to individual and collective ruin.[7] For us, sufficiency represents a common thread linking affluence, technology, and population while addressing the ecological overshoot and socio-economic inequality of our times. Accordingly, we suggest that the

5 A. Hornborg, "Zero-Sum World," *Int. J. of Comparative Sociology* 50, n. 3–4 (2009); C. Dorninger, A. Hornborg, D. J. Abson et al., "Global patterns of ecologically unequal exchange," *Ecological Economics*, 179 (2021); E. Pineault, *A Social Ecology of Capital* (Pluto Press, 2023).

6 A. Malm and the Zetkin Collective, *White Skin, Black Fuel: On the Danger of Fossil Fascism* (2021, Verso).

7 E.g., T. Princen, "Principles for Sustainability," *Global Environmental Policy* 3, no. 1 (2003); T. Princen, *The Logic of Sufficiency* (MIT Press, 2005).

world should have *sufficient* (not too much, not too little) affluence, *sufficient* (not too much, and not too resource-intensive) technology, and a *sufficient* (not too big, not too little) population. Compelling arguments can be made that, for each of the three aspects, we only need that which is necessary for a flourishing and fulfilling life. We perceive that such a sufficiency approach has the potential to meet essential human needs in a just and fair manner while taking into account that we must also consider the well-being of non-human others.[8]

Hoping to offer novel insights into sustainability debates, we have asked the contributors of this book to treat the A, T, and P factors critically by focusing on a single component of the I = PAT equation and without shying away from difficult and sensitive matters. To be sure, it is important to acknowledge that—because of their sensitivity and the historical, ongoing, and possible future exploitation of the Global South by the North—one should approach affluence, technology, and population in a careful and nuanced manner. In practice, this means that the contributors of this book should be aware of their personal socio-economic status and privilege.

To further elaborate on the contents of the book, we examine the concept and ethos of sufficiency before turning to providing a short overview of sufficiency in regard to affluence, technology, and population.

1 Sufficiency as a Concept and Ethos

In recent decades, a key policy strategy to reduce environmental impact has been to focus on improved *efficiency*. Consequently, wide-ranging attempts have been made to increase the efficiency of various processes of manufacturing, distribution, recycling, and so forth in order to reduce emissions and resource input per unit.[9] Despite increased efficiencies and the promise that those efficiencies would be taken to lessen environmental burdens, they have actually supported economic growth and the attendant increase in matter-energy throughput and emissions.[10] In the contemporary economic growth-driven world economy, this increase occurs because efficiency gains typically reduce costs and increase profit margins, which lead to monetary surpluses that are then directed toward further increasing production and consumption—a

8 E.g., K, Bonnedahl and P. Heikkurinen, *Strongly Sustainable Societies* (Earthscan, 2019).

9 B. Alcott, "The Sufficiency Strategy," *Ecological Economics* 64, no. 4 (2008); P. Heindl and P. Kanschik, "Ecological Sufficiency," *Ecological Economics* 126 (2016).

10 See, e.g., J. Hickel and G. Kallis, "Is Green Growth Possible?" *New Political Economy* 25, no. 4 (2019); Pineault (2023).

phenomenon known as the "rebound effect" or Jevons' paradox.[11] As a result, efficiency seems to be an ineffective policy target, from a sustainability perspective, as long as economic growth continues and repeatedly cancels out the achieved efficiency gains.

In response to this situation, scholars and activists have begun to pursue alternative strategies, ones that would result in absolute reductions of the aggregate matter-energy throughput of the global economy and thus reduce overall environmental impacts. One promising alternative strategy is to consider the concept and ethos of *sufficiency*.[12] In the current state of overshoot, there evidently exists the need for measures that would lead to absolute reductions in matter-energy throughput. We argue that sufficiency, as an ethos, has the potential to do just that; this is true because it not only addresses high-consuming population segments,[13] it is also equally concerned with societal structures, social justice, and governmental policies. Furthermore, in this edited volume, we argue that sufficiency, as a concept and ethos, is applicable to affluence, technology, and population.

Sufficiency has been described as both a means and an end.[14] This is because the ethos of sufficiency contains the notion of what can colloquially be called *enoughness*. At the same time, *sufficiency* can imply quantitative thresholds, which apply to, for example, ecological carrying capacity (the maximal number of persons or per capita consumption) or sufficient affluence (the minimal levels of income required to meet basic needs). This further suggests an appropriate "space" between maxima and minima where human needs are met within the carrying capacity of the planet; this space has also been referred to as a "consumption corridor" or as the "doughnut" model.[15] The need to stay within the consumption corridor entails consequences for the A, T, and P factors, as discussed in this book. At the same time, we perceive that sufficiency's greatest change potential is in the ethos of enoughness and in the concept serving as an organizing principle, even an ethic. A central question is then how to transition away from overshoot and growth, guided by an ethos of sufficiency, in a just and planned fashion; as we will see, the journey is hazardous and untested, precisely because it has, as of yet, not been attempted.

---

11 Alcott (2008); A. Berner et al., "Do Efficiency Improvements Reduce Energy Use?" *Energy Economics* 110 (2022).

12 See, e.g., Princen (2003); (2005).

13 M. Sandberg, "Sufficiency Transitions," *Journal of Cleaner Production* 293 (2021).

14 J. Jungell-Michelsson and P. Heikkurinen, "Sufficiency," *Ecological Economics* 195 (2022).

15 D. Fuchs et al. *Consumption Corridors* (Routledge, 2021); K. Raworth, *Doughnut Economics* (Random House, 2017).

One way to grasp the current situation while also acknowledging the human impact on the biosphere is to consider environmental footprint calculations. At present, there is a wide range of environmental footprints available, including ones for carbon, materials, water, and chemicals. While each approach has its disadvantages (such as being centered on the individual as opposed to being centered on societal structures), we suggest that they provide important evidence, for instance, information on the scale of change required. Additionally, they can make visible the differences between the environmental impact of different lifestyle and income classes because (as already mentioned) not all humans contribute (nor have they contributed) equally to the ongoing ecological crisis.

Let us take the carbon footprint as an example. To limit global warming to 1.5 degrees Celsius, total global carbon emissions should not exceed 20 billion tons annually, or about 2.5 tons per capita, on average, by 2030.[16] But of course, there are huge disparities. Currently, the most affluent 10% of the human population emit half of all global greenhouse gas emissions and have an average per capita footprint of 18 tons.[17] On a national level, China and the US lead the world at around 10 billion tons/year and 5 billion tons/year, respectively. The citizens of some small countries top the list as per capita emitters: Qatar, Montenegro, and Kuwait (all emitting over 25 tons per capita).

In social and societal contexts, sufficiency is not a fixed framework or target but something that is culture and context dependent. Hence, the minimum threshold of the consumption corridor, or the social foundation of the "doughnut," is a matter of debate (and rightly so). If reaching agreement on this seems like a great challenge, it is worth remembering that humans always negotiate with each other on proposed actions to be taken.

However, things are different in any *ecological* context. It is not possible to negotiate with, or to cover up, the biophysical realities of the planet or a specific bioregion; thus, ecological sustainability is non-negotiable. These two factors—social justice and ecological sustainability—understandably complicate the debate concerning sustainability transitions and responsibilities.

Therefore, sufficiency is about quality and meaning as much it is about asking "How much does a nation, or a community, or a person need?" or "How much is needed *for a good life?*" To these ends, it can be helpful to distinguish between wants and needs, between luxuries and necessities. In a world of sufficiency, private luxury needs to give way to necessity in order to attain a good

16 L. Akenji et al., *1.5-Degree Lifestyles* (Hot or Cool Institute, 2021).

17 A. Khalfan et al., "*Climate Equality: A Planet for the 99%* (Oxfam 2023); Oxfam, *Extreme Carbon Inequality* (Oxfam, 2015).

life for all (to the degree that that is possible) while staying within local and global ecological carrying capacities. In this sense, sufficiency challenges the dominant logic of growth-driven economics and economies, of growth-based politics, and of societies building on the ethos of an ever-increasing standard of living and the endless accumulation of capital—all of which take a toll on both human and non-human life.

For all this, the answers to the questions of what a "good life within planetary boundaries"[18] looks like and what kinds of actions are needed to attain it, remain unclear. To begin to address this lack of clarity, the contributions in this book converge on at least one point: a societal change toward sustainability is required, especially in the overconsuming economies, and this can be achieved by focusing on sufficiency in regard to affluence, technology, and population.

Correspondingly, each of the individual factors (A, T & P) have dedicated parts in this book, and thus, in the remainder of this Introduction we outline the main themes, structure, and content of this edited volume.

2 Sufficiency in Regard to Affluence, Technology, and Population

Perhaps a first step toward sufficiency is to acknowledge that a book like this is something of a paradox. On the one hand, the contributors are typical representatives of contemporary academic discourse in the West: they are generally middle-class, of mixed genders, and perhaps most importantly, are all members of affluent, technologically advanced nations. On the other hand, it is precisely such people and such societies that, by and large, have created the planetary ecological crisis. Even if none of us have much deepened the ecological overshoot, it is nonetheless true that we participate in the processes that furthers climate change, ecological destruction, and mass extinction. Hence, the contributors are immediately placed in an awkward situation; the very people "guilty" of causing the crisis now situate themselves as the leading intellectuals in a movement to solve the crisis.

Therefore, we, the contributors, should acknowledge this fact, even as we attempt to find our way out of the abyss. The ecological overshoot has emerged over many decades, with centuries-old roots, and yet, any solutions must be forward looking. To be sure, the purpose of this book is not to cast blame either but to use the available means to envision and articulate positive, humane,

18 D. O'Neill et al., "A Good Life for All within Planetary Boundaries," *Nature Sustainability* 1, no. 2 (2018).

just, and sufficient ways forward. In this sense, the contributors are in a good position to analyze and judge such ways because, first, they have the requisite training and expertise and, second, it is largely the overly affluent and technological societies that must change what they are doing and rethink their business as usual.

2.1 *Affluence: Degrowth for the Overshooters, Empowerment for Those Struggling*

Ecological overshoot indicates that humans use, shape, destroy, and pollute too much nature. The Anthropocene, as the proposed new geological epoch, is an important concept,[19] but it tells us very little about who is responsible for global environmental degradation or what kinds of actions, processes, and structures are responsible for the degradation. On the one hand, it is obvious that high-consuming individuals, organizations, and societies bear a disproportionate burden for ecological overshoot, but on the other hand, we must address the aspiration for wealth found among other classes of people.

From the perspective of sustainability and sufficiency, we argue that there is such a thing as "too much affluence."[20] To transition from overshoot to enoughness, there is a call for an overall decrease of matter-energy throughput, coupled with a fair and planned redistribution of wealth and resources for those in need. In short, there must be *degrowth* for the overshooters and *empowerment* for those struggling.[21]

This sounds simple, but surely it is not. The dominant economic organization of the world, capitalism, is based on the expansive accumulation of capital, entailing the push for perpetual economic growth, growth-based politics, and growth-dependent societal structures.[22] Because capital translates to power, the richest can steer politics, shape public opinion by controlling the media, build walls around themselves, hire armies to fight for them, and of course, blame those who are lacking and vulnerable for all the problems in the world. Moreover, the richest people are not likely to give up what they have willingly. The only way that degrowth politics and policies have a chance is if they have enough public support, but at the moment, the majority of people think that their personal situation is going to improve if the economy grows. In opposing

19 See, e.g., P. Heikkurinen et al., "The Anthropocene Exit," *Ecological Economics* 164 (2019).

20 E.g., I. Robeyns, "Why Limitarianism?," *The Journal of Political Philosophy* 30(2), 249–270 (2022).

21 See, e.g., G. Kallis et al., *The Case for Degrowth* (Polity, 2020); J. Hickel, *Less is More* (Penguin Random House, 2021); and M. Schmeltzer et al., *The Future is Degrowth* (Verso, 2022).

22 K. Saito, *Marx in the Anthropocene,* (Cambridge University Press, 2023); K. Saito, *Slow Down,* (Weidenfeld & Nicolson, 2024).

corners, we have short-term personal and societal gains and the long-term viability of the planet. Culturally, it is difficult to ask people to give up their growth-enabled dreams, identities, and careers, even if they are reminded that, after all, they come at a huge cost and, in any case, economic growth tends to typically benefit a small and privileged minority the most. Anyhow, by calling for redistribution, relocalization, and deep transformations, we think it is critical to envision a world of "sufficient affluence" with stable livelihoods and diverse, community-based economies, where each member gets his or her just share without acquiring too much at the expense of others.

2.2 *Technology: Abandoning the Silver-Bullet Metaphor*

Taking action to address affluence, technology, and population in a just manner is nearly impossible in the present political climate for any individual nation, let alone for humanity as a whole. Under the current political conditions, there is neither the vision nor the will to make the necessary changes. Yet, we know from history that national and global situations can change rapidly and unexpectedly; hence, the door for change is always slightly ajar, especially in such a turbulent geopolitical moment as ours and amidst global environmental degradation. To be sure, the overly affluent are not likely to surrender their fortunes, namely because they feel entitled to their possessions and because they are protected by legislation and state power.

Although affluence and technology are deeply intertwined in the unequal world economy, technology remains a different phenomenon. This is because technology is widely considered to be both a *problem* and a *solution* at the same time. Whereas it is counterintuitive to say that the problem of too much affluence will be solved by making everyone rich, it is peculiar that many hold the opinion that the problems of technology will be alleviated by developing and manufacturing more technology. Indeed, technological optimism is firmly entrenched in many people, causing them to dismiss the many contemporary problems regarding the design, manufacturing, development, and use of technology.

Technology optimists believe that, even if technology in not currently at the desired state, it will—sooner or later—get better, healthier, and cleaner. Unfortunately, technology consistently fails to fulfill the expectations. First, the solutions are often *ad hoc*, temporary bandages that only solve immediate effects at the expense of longer-term root causes. Second, the solutions often have unanticipated side-effects that can be as bad, or worse, as the original problem(s). Third, tech solutions are often unjust; they benefit the wealthy and powerful at the expense of the majority. Fourth, such solutions, in failing to address the underlying technological causes, allow extractive and exploitive tendencies to continue or even to accelerate.

Moreover, industrialism and perpetual economic growth have been permitted by both the discovery of and expansive utilization of fossil energy. Without such energy, there would not be an industrial age, an ecological crisis, transnational ultra-rich elites, or the geology of humankind. However, within perhaps a half century, as fossil fuel use inevitably declines, humanity will return to a non-fossil-fuel state with industrial infrastructure and advanced technological know-how. Whether "sustainable energy" sources and transitional technologies will make up for the deep fossil descent remains to be seen.

Technology always has a material footprint, and the more complex and resource-intensive a technology is, the bigger the footprint is. Technological "solutions" seem to increase the footprint of both of the subject technology itself and the larger support system in which it is embedded. One can thus argue that the design, manufacturing, development, and use of technology is always a contested and paradoxical phenomenon and, hence, that the silver-bullet metaphor of technology should be abandoned.[23]

What, then, is "sufficient technology"? We propose the following: it is the minimum level of technology that meets a need or serves a purpose without adding to social conflict and environmental degradation. It refers to ways of being and practices that empower and ease the lives of its users without doing so at the expense of others (including the rest of nature). This description seldom goes together with modern, complex, and resource-intensive technologies like nuclear power plants, electric cars, wind turbines, or solar energy systems; but it is significantly more compatible with "low and slow" technological solutions like bicycles, carts, hand tools, and handicrafts.[24]

2.3 *Population: How to Deal with the Population Question?*

The history of the population debate is ugly, to say the least. To consider its worst aspects, it can contain racism, xenophobia, traces of supremacy, ethnonationalism, and imperialist oppression. But even if we could brush aside the historical baggage, the population question would remain a delicate topic due to potential restrictions on personal freedoms, human rights, religious ideals, and cultural traditions.

That said, there have never been eight billion people on this planet, and although fertility rates are declining around the world, we still can expect one to two billion additional humans (depending on the estimate) by the end of this century. Therefore, and in the light of too much affluence and technology,

23 See P. Heikkurinen and T. Ruuska, eds., *Sustainability Beyond Technology* (Oxford University Press, 2021).

24 See, e.g., P. Bihouix, *The Age of Low Tech* (Bristol University Press, 2020).

we must also acknowledge that the population factor counts for sustainability. As for any other species on this planet, there are limits to the human population, even if these vary contextually and regionally. In any case, we should not talk about population separately from the material standard of living and the level of technology, but seek out to develop a balanced approach, which is something that this book as whole is trying to do. Environmental impact is, after all, the impact of a number of people utilizing resources at a certain level. From the point of view of the overshoot and carrying capacity, it is the "total footprint" that matters, but we should always remain sensitive to how this footprint is divided among people and to historical and contextual circumstances.

None of these observations necessarily deem that the population question is colonial, racist, or discriminating. Given the global disparities in affluence and technology, incremental population growth in wealthy countries is ecologically more problematic than the equivalent growth in less affluent countries. This must be acknowledged. Yet, we should also take into account that every newborn child requires space, needs (roughly equal) food, has material needs, and more broadly, has an equal claim for having a good life. All these issues are complicating factors in any population discussion, which also should inform us that the problem is global, and one that cannot be fixed by only focusing on human numbers. Further, we should consider that the planet is a shared space with limited resources, and humans are only one species among many. Although the overshooters have a many-times-bigger impact on the earth, every one of us, humans and non-humans, also needs space and resources to live good and flourishing lives.

Much of this is standard fare in contemporary academic debates on population, which fortunately have been able to transcend many of the problems of mainstream politics and social commentaries regarding the conscious and just means with which to downsize human numbers.[25] Thus, the question of sufficiency in population is arguably about achieving a balance by just and fair means where human numbers and well-being correspond to planetary boundaries, where each human gets enough to live a good and fulfilling life, and where the flourishing of diverse non-human life is preserved.

3 The Aims and Structure of This Book

The aim of this edited volume is to enrich sustainability debates, theory, and practice by approaching affluence, technology, and population from the

---

25 E. Cripps, "Population Ethics for an Imperfect World," *The Journal of Development Studies* 57, no. 9 (2021).

point of view of sufficiency. The book asks, and tries to answer, the following questions: What is "sufficiency" in an era defined by ecological overshoot and growth-driven economics and politics? In what ways can reduced affluence—such as affluence reduced through degrowth strategies—be compatible with a good, or even enhanced, quality of life and with various sustainability transitions? How should societies address the issue of relentless technological development and deployment, which threatens to exceed our control and to inflict unprecedented damage upon the natural world? How can we discuss questions related to population in a constructive and just manner? Is it possible to achieve a sustainable and stable human number that leaves room for other species to thrive and flourish? The book also assesses the I = PAT equation itself, considering recent advancements in natural and social sciences, economics, and environmental ethics.

The reader should also note that each author offers their point of view to the issue at hand, indicating several perspectives that may or may not always be compatible with each other. Furthermore, the aim is not simply to conduct an analysis, but also to generate prescription and policy recommendations; the contributors to the book all strive to construct realistic and actionable policies for all the three factors using the sufficiency concept and ethos as a guiding principle. These policy suggestions target the entire social hierarchy, ranging from national and international recommendations to state, regional, local, and consumer-level actions.

Regarding the organization of the book, our rearranged equation, I = ATP, provides a template by which to structure the book. The book has four parts: "Setting the Stage" (Part 1), to which this introduction belongs, opens the book with a chapter on relational sufficiency and a discussion on the missing activities in the I = PAT equation.

"Sufficiency in Affluence" (Part 2) deals with the need to curtail the economics of perpetual growth, namely capitalism and others forms of productivism, to implement absolute reductions in economic throughput and to create fair and equitable economic outcomes for all.

"Sufficiency in Technology" (Part 3) examines the role and effects of modern resource-intensive technology as it affects social structures, transportation systems, and the human psyche. The notion of technological sufficiency is seen in a new light in all these areas, in contrast to apparently endless technological acceleration.

"Sufficiency in Population" (Part 4) then discusses the many dilemmas linked to population. These issues are analyzed in the light of ecological questions and environmental metrics, such as the ecological footprint, and are considered from various social justice-related standpoints.

4 Chapter Outlines

In Part 1, "Setting the Stage," Chapter 2, one of the defining scholars of the sufficiency concept in its contemporary understanding, Thomas Princen, outlines sufficiency as an imperative used to address the urgent problem of *excess*. He examines sufficiency as both a personal idea—the sense of enoughness and "too-muchness" that leads to a practice of the right amount—and as a social organizing principle for doing well both now and in the distant future. The sufficiency that Princen posits concerns both social and ecological dimensions and is primarily relational, participating in a way that does not deprive others. In Chapter 3, Mikko Jalas discusses the missing activities in the I = PAT equation. He calls for new and qualitatively different understandings of organizing human–nature relations. Jalas claims that this requires rethinking both affluence and technology, and does so by relating them to *time*. Affluence is redefined as wealth in time, *time affluence* is measured as one's hours outside of paid work, and technology is considered in terms of the resource intensity of everyday activity. A new formulation of the equation is proposed: I = *PACT*, where the factor C is introduced to the equation signifying consumption.

In Part 2, "Sufficiency in Affluence," in Chapter 4, Tina Nyfors and Lassi Linnanen aim at operationalizing sufficiency by linking income classes with environmental footprints, considered in relation to planetary boundaries. Particular attention is given to the most carbon-intensive options, footprint hotspots. Based on calculations, they propose steps for making sufficiency-based proposals for seven lifestyle classes, highlighting the fact that a new role model—the "sufficiency class"—already exists. In Chapter 5, Toni Ruuska portrays how the Global North takes advantage of subordinate peoples, landscapes, and habitats to maintain its privileged position in the global capitalist economy. Instead of there being more affluence at the expense of others, the chapter calls for degrowth and the redistribution of wealth and resources. Affluence and technology are discussed from a sufficiency perspective to find pathways towards sustainability. In Chapter 6, Iana Nesterova and Jessica Jungell-Michelsson focus on businesses as important actors in deep transformations. The chapter defines *degrowth* as deviating from the pursuit of economic growth for its own sake, but it does not equate it with a reduction in all economic and business activities. Rather, it highlights the need for growth in certain industries that satisfy human needs while manifesting gentleness and care towards nature. It proposes that the degrowth business is a business of sufficiency.

In Part 3, "Sufficiency in Technology," in Chapter 7, Tere Vadén discusses technology, affluence, and the role of fossil fuels, with its huge energy density, in the perceived mastery of humanity over nature. The chapter examines what Vaden calls *nafthism*—the masking of the crucial role of fossil energy. The chapter argues that in fossil capitalism, affluence, technology, and population are elementally dependent on fossil fuels. It suggests an urgent double task: first, to reappraise the modern experience of technology in the "black light of oil" and, second, taking that reappraisal seriously with the approaching end of the fossil fuel era. In Chapter 8, Karl Johan Bonnedahl calls for looking at sustainability and development in new ways and discusses how a different approach to technology may provide a pathway to sustainability. Bonnedahl identifies continuous economic growth and unlimited individual demand as the reasons behind unsustainability and discusses an alternative approach to technology where social innovation and increased attention to well-being and needs are key. In Chapter 9, J. Mohorčich considers transportation in a future, energy-constrained world. Six modes of transportation are analyzed under two scenarios: one in which humans have 50% less energy to use for transportation and one in which they have 95% less. The chapter finds that the implications of any post-Anthropocene scenario on transportation are considerable, and that transportation mode and speed are decisive factors for future mobility.

In the final part, Part 4, "Sufficiency in Population," in Chapter 10, Michel Bourban discusses the ethical and political implications of the planetary boundary framework, with a particular focus on the effects of population on climate change and biodiversity loss. This chapter explores and evaluates population policies and proposes possible justifications for certain positive and negative incentive measures. This chapter also considers objections commonly made against these incentive policies. In Chapter 11, Pasi Heikkurinen and Joonas Uotinen raise the question of how to talk about the unsustainability of population growth in affluent societies (mainly located in the northern hemisphere). Also, they discuss whether it is meaningful to talk about a "sufficient population." The chapter both acknowledges the sensitivity of the topic and considers it problematic if population is left outside the sustainability debate. It suggests a five-point checklist for a more appropriate discourse on a sufficient population. In Chapter 12, Phil Cafaro discusses applying the I = PAT equation to an overpopulated world and explores why discussing population is so hard. This is exemplified by how population was dealt with in the IPCC mitigation report of 2022. In it, population growth was dealt with as an important driver of increased greenhouse gas emissions, but it was excluded from the executive summary for policymakers. The chapter concludes by suggesting policies to promote substantially and justly reduce the current human population.

In the concluding chapter, Chapter 13, Tina Nyfors and Toni Ruuska discuss sufficiency-driven change and summarize the key findings from each part of the book.

Acknowledgements

This work was supported by funding from the Research Council of Finland's project *"The Skills of Self-provisioning in Rural Communities"*, grant number 343277, the Kone Foundation's project *"The Underdogs of Just Transition"* (*Oikeudenmukaisen siirtymän altavastaajat*), grant number 202302807, and the Tiina and Antti Herlin Foundation.

References

Akenji, L. et al. (2021). *1.5-Degree Lifestyles: Towards A Fair Consumption Space for All.* Hot or Cool Institute.

Alcott, B. (2005). "Jevons' Paradox." *Ecological Economics* 54: 9–21.

Alcott, B. (2008). "The Sufficiency Strategy: Would Rich-World Frugality Lower Environmental Impact?" *Ecological Economics* 64, no. 4 (February): 770–86.

Berner, A. et al. (2022). "Do Energy Efficiency Improvements Reduce Energy Use? Empirical Evidence on the Economy-Wide Rebound Effect in Europe and the United States." *Energy Economics* 110, 105939.

Bihouix, P. (2020). *The Age of Low Tech: Towards a Technologically Sustainable Civilization. Bristol: Bristol University Press.*

Bonnedahl, K., and Heikkurinen, P. (2019). *Strongly Sustainable Societies: Organizing Human Activities on a Hot and Full Earth.* Earthscan.

Chertow, M. (2008). "The IPAT and its Variants." *Journal of Industrial Ecology* 4, no. 4: 13–29.

Commoner, B. (1971). *The Closing Circle. Nature, Man, and Technology.* Alfred A. Knopf.

Cripps, E. (2021). "Population Ethics for an Imperfect World: Basic Justice, Reasonable Disagreement, and Unavoidable Value Judgements." *The Journal of Development Studies* 57, no. 9: 1470–1482.

Dorninger C., Hornborg, A., Abson, D., et al. (2021). "Global patterns of ecologically unequal exchange: Implications for sustainability in the 21st century," *Ecological Economics*, 179: 1–14.

Ehrlich, P. and Holdren, J. (1972). "Critique." *Bulletin of the Atomic Scientists* 28, no. 5: 16–27.

Fuchs, D. et al. (2021). *Consumption Corridors: Living a Good Life within Sustainable Limits.* Oxon, New York: Routledge.

Heikkurinen P., and Ruuska T., eds. (2021). *Sustainability Beyond Technology: Philosophy, Critique, and Implications for Human Organization.* Oxford University Press.

Heikkurinen, P. et al. (2019). "The Anthropocene Exit: Reconciling Discursive Tensions on the new geological epoch." *Ecological Economics* 164, 106369.

Heindl, P., and Kanschik, P. (2016). "Ecological Sufficiency, Individual Liberties, and Distributive Justice: Implications for Policy Making." *Ecological Economics* 126: 42–50.

Hickel, J. (2021). *Less is More: How Degrowth Will Save the World.* Penguin Random House.

Hickel, J. and Kallis, G. (2019). "Is Green Growth Possible?" *New Political Economy* 25, no. 4: 469–486.

Hornborg, A. (2009). "Zero-Sum World: Challenges in Conceptualizing Environmental Load Displacement and Ecologically Unequal Exchange in the World-System," *International Journal of Comparative Sociology*, 50(3–4): 237–262.

Jungell-Michelsson, J. and Heikkurinen, P. (2022). "Sufficiency: A Systematic Literature Review." *Ecological Economics* 195: 1–13.

Kallis, G., Paulson, S., D'Alisa, G. and Demaria, F. (2020). *The Case for Degrowth.* Polity.

Khalfan, A. et al. (2023). *Climate Equality: A Planet for the 99%.* Oxfam International.

Malm, A. and the Zetkin Collective (2021). *White Skin, Black Fuel: On the Danger of Fossil Fascism.* Verso.

O'Neill, D. et al. (2018). "A Good Life for All within Planetary Boundaries." *Nature Sustainability* 1, no. 2: 88–95.

Oxfam (2015). *Extreme Carbon Inequality.* Oxfam.

Pineault, E. (2023). *A Social Ecology of Capital.* Pluto Press.

Princen, T. (2003). "Principles for Sustainability: From Cooperation and Efficiency to Sufficiency." *Global Environmental Politics* 3, no. 1: 33–50.

Princen, T. (2005). *The Logic of Sufficiency.* MIT Press.

Raworth, K. (2017). *Doughnut Economics: Seven Ways to Think Like a 21st-Century Economist.* Random House.

Rees, W. (2020). "Ecological Economics for Humanity's Plaque Phase." *Ecological Economics* 169, 106510

Richardson, J. et.al. (2023). "Earth beyond six of nine planetary boundaries." *Science Advances* 9: 37.

Robeyns, I. (2022). "Why Limitarianism?," *The Journal of Political Philosophy,* 30(2), 249–270.

Saito, K. (2023). *Marx in the Anthropocene: Towards the Idea of Degrowth Communism.* Cambridge University Press.

Saito, K. (2024). Slow Down: How Degrowth Communism Can Save the Earth. Weidenfeld & Nicolson.

Sandberg, M. (2021). "Sufficiency Transitions: A Review of Consumption Changes for Environmental Sustainability." *Journal of Cleaner Production* 293, 126097.
Schmeltzer, M., Vetter, A. and Vansintjan, A. (2022). *The Future is Degrowth: A Guide to a World beyond Capitalism*. Verso.
Steffen, W., Broadgate W., Deutsch L., et al. (2015). "The trajectory of the Anthropocene: The great acceleration," *Anthropocene Review*, 2(1): 81–98.
York, R., Rosa, E. and Dietz, T. (2003). "STIRPAT, IPAT and ImPACT: analytic tools for unpacking the driving forces of environmental impacts." *Ecological Economics* 46, no. 3: 351–365.

CHAPTER 2

Relational Sufficiency: an Essay on an Idea, a Principle, an Imperative

Thomas Princen

1 An Idea, a Principle

In the interests of full disclosure right up front, here is what I mean by 'sufficiency,' put simply and concisely:

> *As a personal idea*: Sufficiency is that sense of enoughness and toomuchness that leads to a *practice* of the right amount.
> *As a social organizing principle:* Sufficiency is doing well now and into the distant future, by *organizing* to do less than the most possible now.

Put differently, like environmental thought generally, sufficiency can be posed in relational terms or impact terms. In Sweden (I have observed) the term *lagom* ("enough" or "just the right amount") comes from a beer-drinking ritual. We're sitting around a table drinking beer, and I ask, Is it sufficient? That is, is the amount I have poured in my mug enough? If I glance at others' mugs, they all have enough to drink and the pitcher's nearly empty, what I'm taking is enough but not too much, just right, sufficient.

By contrast, if I ask a land manager or an international development aid officer working to relieve poverty if it is sufficient, "it" being a harvest rate or a funding program, the individual can answer only after doing some calculations. The land manager needs data to be sure the harvest is below or equal to regrowth. The development officer needs a measure of basic income or basic needs to know if the funding level is enough.

The difference is between a relational and impact understanding of sufficiency. The beer drinking principle of sufficiency can be generalized (as it is in Sweden) to a host of collective decisions, including policy decisions. To ask if an action is sufficient is to put me, the individual, or the town or the company or the political party, into a context of a larger whole: am I taking my share, enough to situate me in this group, to be a member, to do my part, to be "one of the folks," to belong and do so without deleteriously affecting others in the group? Am I getting what I need for myself, including what I need to be a bona

 | DOI:10.1163/9789004746404_003

fide member, and leaving enough for everyone else to do likewise? In other words, relational sufficiency is participating in a way that does not deprive others, does not elevate some at the expense of others and is highly attentive to taking too much, that is, to excess. It signals our collective membership. Most important, perhaps, it *elevates the group as a whole*. And, then, on the biophysical side, relational sufficiency situates the group within a biophysical environment, *using natural resources without using them up*.

The land and aid examples, by contrast, focus on metrics, on quantifiable levels of extraction and consumption all of which can be managed and from which, presumably, effective social relations follow. The beer-drinking example embeds relations in the very act of taking enough and not too much. The mechanistic, managerial approach abstracts away context and relations to focus on the measure—board feet and money, say. This approach is optimizing and efficiency seeking, its time frame the immediate term. It presumes that other important things in life—personal relations, group cohesion, meaning in work and play, a sense of wonder, living with natural systems, not dominating them—will follow. In a sense, it presumes to be laying a foundation, to being a form of grounding. But its implicit world view is mechanistic, not relational. It perceives the world atomistically, not holistically. It conceptualizes the world as a machine, not an organism or ecosystem. It values that which can be measured, discounting, however inadvertently, all else. It employs the language of the factory. What is more, it presumes an elite to manage it all and an underclass to take orders.[1]

The sufficiency posited here then is, in the first instance, *relational*, only in the second measurable or quantitative or managerial.[2] It aspires to be a social organizing principle on the order of consumer sovereignty, efficiency, and growth, the dominant social organizing principles of modern, industrial, consumerist, commercialist societies. Its relationality is both *social* and *ecological*. It thus complements the impact focus of IPAT. What is more, the defining context is a world of *excess*—the exceeding of biophysical capacities, social organizing capacities, and psychological, attentional capacities—facilitated by a language of extraction, expansion, and externalization. My premise is that, because humanity is in for fundamental cultural change, incremental adjustment is inadequate. We need concepts for such change, indeed, a language—principles, metaphors, stories—that counters excess, that embodies the needs

1 On the relationship between sufficiency and social structure, see Princen T (2022) Sufficiency and the state: A prospective project. *Front. Sustain.* 3:956139. doi: 10.3389/frsus.2022.956139.

2 On the political difference between the relational and managerial, see M. Walzer, *Spheres of Justice* (Basic Books, 1983).

of the 21st century, not the industrial, expansionist, consumerist needs of the 19th and 20th centuries. Sufficiency is one element of that language. It is most useful when aimed at the overconsuming Global North.[3] What is more, it is most useful when limited to, or at least applied primarily to relations with the material world, especially natural resources.

2 A Transitional Principle

So if sufficiency isn't in the first instance about measurement and optimization, let alone satisfying consumers and promoting economic growth, what is it? What does it aspire to organize society to do? On the environmental front, in the early years of pollution control, it might have been invoked to determine an "optimal level of pollution" or "acceptable damage." As sustainability gained currency from the 1980s onward, sufficiency might have been invoked to guide development in a way that allowed current generations to meet their needs without compromising the needs of future generations. For those who dared break the taboo against criticizing growth, who challenged the pervasive sense of limitlessness, people like systems analyst Donella Meadows, ecological economist Herman Daly, and social theorist Wolfgang Sachs, it might have been invoked to keep the economy within biophysical limits.

But now, in the 21st century, I assert, sufficiency's role is to help guide the transition, what some call transformation, either term implying fundamental, discontinuous, societal shift, not just a swapping out of fossil fuels and swapping in of renewables. Because this transition, as opposed to previous transitions—for example, from hunter-gatherer to settled agriculture, from feudal to industrial, from slave state to free state, from wood fuel to fossil fuel—is necessarily and unavoidably about adapting to declining available resources, human and natural, sufficiency points to both measurable levels and relations. It asks not just how much is enough and too much, prompting for some an exercise in measurement and implying that people can continue to choose

3 I use Global North and Global South not as geographic terms but as designations of wealth and consumption and, ultimately, economic and political power. At a country level, the Global North includes the G20 countries, the Global South the LDCs, so-called Less Developed Countries. More significantly, though, the Global North can be found in, say, the capital cities of poor countries (Makati district of Manila, for instance) and the Global South in marginalized communities of rich countries (Flint, Michigan in the United States, for instance). Regarding IPAT, the Global North has the greatest impact largely as a function of its consumption (affluence) and technologies (especially of extraction, distribution, and deposition).

from an ever-expanding basket of goods. It presumes that some will downsize out of concern for the environment or the poor, while others will consume up to their budget constraint (and often beyond). On the material side, it asks how much is available now, and how much is likely to be available in the future. As for social relations—familial and neighborly, commercial, and spiritual—and social structures—it asks what rules and procedures, norms and principles—are compatible with declining available resources. As for ecological relations, it asks for *congruence* between social and ecological organizing principles.[4]

Material flow analysts, along with their econometric and engineering brethren, will be well employed addressing the measurement questions. They will, I surmise, have little trouble shifting from their current preoccupation of how to grow the economy, how to disrupt established means of communicating and selling, how to lay more infrastructure, how to maximize return on investment, how to market a product in novel ways, and so forth, to estimating constraints and devising formulas for distribution. They will have trouble, though, if they are called on to reorganize society for resource descent, a defining feature of the transition.[5] Here there will be a great clash of cultures. Their organizational and institutional culture, being one of growth, limitlessness, sales, jobs, and the like, will collide with those who see limits everywhere, not just as constraints but as expressions of being human, of being in a place, of living well and well within the means of single ecosystems, a single hydrologic system, a single climate system.

3 Why Sufficiency?

Why explore sufficiency? For one, it is underexplored, certainly in my field—global environmental politics—a subfield of international relations where primary concerns are for international cooperation, organization, institutions, and negotiations. Second, sufficiency and like terms—frugality, thrift, moderation, simple living—have been around for a long time but need updating. But to me the most important reason is the contemporary need, in fact, a desperate social need, for a concept like this. The biophysical trends—often

4 T. Princen, "Sustainability: Foundational Principles," in *Routledge Handbook of Global Environmental Politics*, ed. P. Harris (Routledge, 2014).

5 R. De Young and T. Princen, *The Localization Reader* (MIT Press, 2012); T. Princen, "The Politics of Urgent Transition," in *US Climate Change Policy and Civic Society*, ed. Y. Wolinsky (CQ Press, 2014).

lumped under the banner of "climate change"—not to mention the political, economic, and cultural trends, cry out for

1. a concept other than growth, efficiency, consumer sovereignty, speed, mobility, open borders, anything goes, move fast and break things, disruption;
2. conceptual aids for reorganizing what is fast proving to be an unsustainable and unjust social order;
3. a language that is at once counter to the dominant language of domination and exploitation, efficiency and growth, and consonant with a primary need of the times, namely, living within our biophysical, organizational, and social psychological capacities or, put differently, living *with nature*, organizing in ways consonant with how natural systems organize themselves and how humans are constituted (for, say, belonging, meaning, novelty and familiarity).

Anecdotally, the mere fact that questions of how much is enough and how much too much are, from where I sit (in the US, in a major research university), taboo, suggests the need for a conceptualization of something like sufficiency. So not only is a growing economy, a growing population, and growing consumption presumed good, but in my university (including in a school of sustainability) so is a growing student population, growing faculty, growing space, and, above all, growing research and tuition dollars. To speak otherwise is to invite gasps, disbelief, ridicule, shunning. In short, growth and efficiency are taken as self-evident in the mainstream public discourse I encounter.

The taboo boils down to this: Don't talk about doing less. If there's an environmental problem, clean it up, do things green, be more efficient, grow a green economy, develop new technologies, promote alternative energy sources. But do not talk about doing less. Consequently, much of my work has aimed at breaking that taboo, at legitimizing ways of living, including producing and consuming as if we have only one planet, as if we can do well by doing enough but not too much. Of course, "less" is not sufficiency per se. But for the overconsuming Global North it is a step towards an appropriate, sustainable, just scale of human throughput.

4 Purpose

So the development of such an idea, concept, principle should serve a purpose. For me, that purpose ought to be more than academic. In other words, there is a politics in such an exercise. The intellectually honest thing is to be explicit about it. From my vantage point (indeed, my advantage point) I see a world of *excess*—the exceeding of biophysical capacities, the exceeding of social organizing capacities, and the exceeding of psychological, attentional capacities—all facilitated by a language of extraction, expansion, and externalization. My purpose is to create a different language, one that goes straight at excess, recognizing that not all actors are implicated equally nor, among those who are implicated, equally responsible. The Global South, for instance, does not carry the same responsibility, historically or politically, as the Global North; consumers, being at the unorganized, low-power end of the chain of decision-making are less responsible than financiers.

A premise is that the organization of the current political economy—call it free market, neoliberal, capitalist, or, my favorite, industrialist/commercialist/consumerist—is a primary determinant of such excess (the other being the state; see below). Because societies organize their resource use—their exploring, extracting, processing, distributing, consuming, and disposing—around social organizing principles, at least implicitly, articulating a counter-principle is in order. In fact, given the trends—from toxics and climate to authoritarianism and pandemics[6]—a counter excess principle is imperative.

Sufficiency, or something like it, is one such principle. And only one. It is not a universal, all-purpose principle (as, e.g., growth and efficiency are often made out to be). Just as sufficiency in practice embraces limits, the concept itself should be limited. For me, in the first instance, sufficiency is most useful, that is it best serves the purpose of organizing against excess, when it is aimed at the overconsuming Global North. What is more, it is most useful when limited to, or at least applied primarily to relations with the material world. And for that, the first application is the most basic, namely, natural resources—water and soil, say—and critical ecosystems—grasslands, forests, estuaries, reefs. A second level application is "producers"—extractors, processors. A third (and here things get murky) is those distant from primary production yet no less determinative—e.g., distributors, financiers. This prioritization of applications follows from a maxim regarding sustainability laid out by the geneticist Wes Jackson who has spent decades developing perennial grains: if society is going to get on a sustainable path it will have to do so first in agriculture. Conversely, it is hard to imagine that a society can be truly sustainable (and just) if buildings and automobiles and mutual funds are "sustainable" and the basic natural resources on which all such systems rest are not.

6 A premise, not developed here, is that the recent pandemic is very much a result of *excessive* human-environment relations, from deforestation and wildlife trade to dense human habitation and cross-border travel.

The material system of relevance here is, in the first instance, a material provisioning system, what may also be called an economy. But it is not just the modern economy that is captured in measures like GDP and interest rates and trade flows. It is the *relational economy* that includes subsistence, household production and reproduction, maintenance of community in place.[7] A material system of sufficiency, a "sufficient economy," is one where decision rules include: Does this action risk exceeding capacities? Is the resulting system throughput enough and not too much? Are buffers, redundancies, and restraint built in? The decision rules under sufficiency supersede efficiency, consumer sovereignty, and growth, the contemporary organizing principles. They precede equity.

One quality of such an economy is risk aversion, not just aversion to high-stakes financial gambles but a sensitivity to high-probability events in the very long term, the ecologically and geologically long term. A certain precarity is thus presumed, biophysical or social or both. Another quality is a social environment where speed, aggression, risk-taking (including risk to others in space or time), expansion are no longer dominant social norms.

I assume that, historically, when big ideas like democracy and peace, growth and sustainability, emerge, they do so to fill a need of their time. They do what other concepts cannot do. They have agents of adoption and diffusion, people with a vision, an interest, an agenda, maybe a grudge. If an idea catches on (serving a need of its time), it tends to agglomerate other ideas; it attracts other interests, agendas, grudges; it comes to express other values and purposes. In the process, it gets bent and stretched. It can lose its usefulness when appropriated for all a sundry things and goals: democracy becomes merely countries that hold elections (however fraudulent); peace becomes nuclear deterrence; efficiency becomes crop yield regardless of soil erosion; growth is the answer to all problems including the problems of growth.[8]

To return then to the question, why sufficiency, my best answer is that it forces one to ask, What is the problem, or What is *the problem*? As noted, for me, it is excess. For others it is inefficiency and waste, insufficient investment, lack of technological innovation. For still others it is poverty and powerlessness,

7 Wendell Berry, *Home Economics* (North Point Press, 1987); Herman E. Daly and John B. Cobb, *For the Common Good* (Beacon Press, 1989); Jack Manno, "Commoditization," in *Confronting Consumption*, ed. T. Princen et al. (MIT Press, 2002); Nelson, J.A. (2011). The Relational Economy. In: Zsolnai, L. (eds) Ethical Principles and Economic Transformation – A Buddhist Approach. Issues in Business Ethics, vol 33. Springer, Dordrecht. https://doi-org.proxy.lib.umich.edu/10.1007/978-90-481-9310-3_2; T. Princen, *Treading Softly* (MIT Press, 2010).

8 For illustration of the bending and stretching of a concept, in this case, efficiency, see "Efficiency: A Brief and Curious History," in T. Princen, *The Logic of Sufficiency* (MIT Press, 2005).

oppression, lack of opportunity. In my ideal world, one intently focused on effecting a positive transition, other principles would be created to meet those problems. Together we'd be moving in the direction of a set of principles, along with metaphors, images and stories, that gets humanity through the 21st century.

In sum, humanity is in for fundamental cultural change, for a transition, like it or not, ready or not. No longer is incremental adjustment adequate. We need concepts for such change, indeed, a language—principles, metaphors, stories—that counters excess, that embodies the needs of the 21st century, not the industrial, expansionist, consumerist needs of the 20th century and earlier. Sufficiency, I submit, is one element of that language. It enables speaking differently.

5 Speaking Differently

I take inspiration from philosopher Richard Rorty's statement that, paraphrasing, fundamental cultural change occurs not when people argue better but when they speak differently.[9] That is, more data, better models, more historical understanding, even better laws and incentives, will not effect the requisite change. The requisite societal change is large scale, long term, and discontinuous, paralleling the biophysical changes now underway. That change, again, I call transition, and is far more than a swap of energy sources.

A part of "speaking differently" is conceptualizing the world and humans' place in it differently, that is, different from endless expansion and consumption, heedless displacement of true costs, gross injustices against the powerless. To my read, current proposals for dealing with climate change, for instance, are in and of the very same language of endless expansion and top-down exploitation: efficiency (carbon tax, carbon trading), cooperation (international agreements like the Montreal Protocol), consumption ("green" consumption), growth (in renewables), technological progress (electrification, carbon capture and sequestration, geoengineering the climate system, next generation nuclear reactors).

Speaking differently means *organizing* ourselves differently, in ways consonant with the ways organisms (including the human organism) and biophysical systems organize themselves. Such systems do not grow endlessly. They build in checks on amplifying, "positive" feedback loops. They are cyclic, not

9 R. Rorty, *Contingency, Irony and Solidarity* (Cambridge University Press, 1989).

linear, intermittent not continuous. They exist as nested systems where no one system (like the modern economy) dominates all others.

Speaking differently means that, collectively, humans relate to natural systems with humility not arrogance and disregard, with restraint not heedlessness, with reverence and awe not hubris, with a positive sense of sacrifice (i.e., make sacred by sacrificing what people *can* do and focusing on what they *should* do).[10]

The project of this transition is one of post-growth, post-consumerism, post-industrialism. It is in part a project of deliberately constructing a language of living within constraints—biophysical, organizational, psychological, ethical. Terms like precaution, prevention, restoration, restraint, Mother Earth, good life (*buen vivir*), kinship are extant nominations for that new language. Sufficiency, I submit, is another.

6 Conclusion

When I started working on the idea of sufficiency in the early 1990s, I thought it was, well, a good idea. Growth and efficiency seem to exacerbate, if not cause, environmental problems. If ecological economist Herman Daly was right, we may find it difficult to define sufficiency, but "it will prove far more difficult to continue to operate [as if] there is no such thing as enough."[11] So I set out to define sufficiency, or better, construct it.

Now I no longer see sufficiency as just a "good idea." I see it, or something like it, or some class of ideas (think moderation, simplicity, frugality, reciprocity, restraint, moral economy, relational economy) as an *imperative*. That is, because all societies organize themselves for natural resource use—for extraction or growing, processing, distributing, consuming, and disposing—and do so, at least implicitly, by employing social organizing principles, the dominant principles must have congruence. Congruence matches societal goals with social capacities and biophysical capacities. For a long time, those principles have been concentration, domination, expansion, and, more recently, efficiency, speed, mobility, open access, consumer sovereignty, and investor sovereignty. All served the state goal of concentrated wealth and power and the state means of surplus and all in a context of virtually endless

10 M. Maniates and J. Myers, eds., *The Environmental Politics of Sacrifice* (MIT Press, 2010); T. Princen, "Toward a Theory of Restraint," *Population and Environment* 18 (1997); G. Van Horn, et al., eds., *Kinship* (Center for Humans and Nature Press, 2021).

11 H. Daly, "The Steady-State Economy" (1993): 361.

resources and waste sinks. The limits of this form of social organization—the state—began to play out financially and politically late 19th century and now is playing out biophysically, ecologically, demographically, epidemiologically.[12]

The 21st century transition thus is not merely swapping out fossil fuels and swapping in renewables, wishing for technological saviors, globalizing further, or localizing anew. It is creating a new social form, one that sees *excess*—exceeding regenerative and absorptive capacities, exceeding social organizing capacities, and exceeding psychological and attentional capacities—as *the problem*. The state as it has developed over some 6,000 years is not, literally, sustainable, let alone just, not on this one planet now full of human impacts as other chapters in this volume have shown via IPAT.[13] Human relations—to each other and to natural systems—are askew. Among the many challenges in the coming decades is righting those relations. One way is to articulate social organizing principles consonant with the human condition on a biophysically constrained planet. Among the nominees, I submit, is sufficiency.

Acknowledgements

I thank Raymond De Young for comments on earlier drafts of this chapter. An earlier version was presented as a paper of the annual meeting of the International Studies Association, March, 2021 (online).

References

Arendt, H. (1948). *The Origins of Totalitarianism*. Harcourt.

Berry, W. (1987) *Home Economics: Fourteen Essays by Wendell Berry*. North Point Press.

Crosby, A. W. (2004) *Ecological Imperialism: The Biological Expansion of Europe, 900–1900. Studies in Environment and History*. Cambridge University Press.

Daly, Herman E. (1993) "The Steady-State Economy: Toward a Political Economy of Biophysical Equilibrium and Moral Growth." In V*aluing the Earth: Economics, Ecology, Ethics*, edited by Herman E. Daly and Kenneth N. Townsend. MIT Press.

12 H. Arendt, *The Origins of Totalitarianism* (Harcourt, 1948); A. W. Crosby, *Ecological Imperialism* (Cambridge University Press, 2004); D. Graeber and D. Wengrow, *The Dawn of Everything* (Farrar, Straus and Giroux, 2021); J. McNeil and P. Engelke, *The Great Acceleration* (Harvard University Press, 2016).

13 For elaboration of how the state form is inherently expansionist and hence antithetical to sufficiency, see Princen "Sufficiency and the State."

Daly, Herman E. and Cobb, J. B. (1989) *For the Common Good: Redirecting the Economy Toward Community, the Environment, and a Sustainable Future*. Beacon Press.

De Young, R. and Princen, T. (2012). *The Localization Reader: Adaptations for the Coming Downshift*. MIT Press.

Graeber, D. and Wengrow, D. (2021). *The Dawn of Everything: A New History of Humanity*. Farrar, Straus and Giroux.

MacNeil, John R. and Engelke, P. (2016). *The Great Acceleration: An Environmental History of the Anthropocene Since 1945*. Harvard University Press.

Maniates, M. and Myers, J. eds. (2010). *The Environmental Politics of Sacrifice*. MIT Press.

Manno, J. (2002). "Commoditization: Consumption Efficiency and an Economy of Care and Connection" In *Confronting Consumption*, edited by T. Princen, M. Maniates, and K. Conca. MIT Press.

Nelson, J.A. (2011). The Relational Economy. In: Zsolnai, L. (eds) Ethical Principles and Economic Transformation – A Buddhist Approach. Issues in Business Ethics, vol 33. Springer, Dordrecht. https://doi-org.proxy.lib.umich.edu/10.1007/978-90-481-9310-3_2

Princen, Thomas. (1997). "Toward a Theory of Restraint." *Population and Environment* 18: 233–254.

Princen, Thomas. (2005). *The Logic of Sufficiency*. MIT Press.

Princen, Thomas. (2010). *Treading Softly: Paths to Ecological Order*. MIT Press.

Princen, Thomas. (2014). "The Politics of Urgent Transition." In *US Climate Change Policy and Civic Society*, edited by Yael Wolinsky. CQ Press.

Princen, Thomas. (2014). "Sustainability: Foundational Principles." In *Routledge Handbook of Global Environmental Politics*, edited by Paul G. Harris. Routledge.

Princen T (2022) Sufficiency and the state: A prospective project. *Front. Sustain.* 3:956139. doi: 10.3389/frsus.2022.956139.Special issue, "From an Ethic of Sufficiency to its Policy and Practice in Late Capitalism," edited by M. V. Mathai, S. Lorek, and W. Sachs.

Rorty, Richard. (1989). *Contingency, Irony and Solidarity*. Cambridge University Press.

Van Horn, G., Kimmerer, R., and Hausdoerffer, J., eds. (2021). *Kinship: Belonging in a World of Relations*, vol. 3, *Partners*. Center for Humans and Nature Press.

Walzer, M. (1983). *Spheres of Justice: A Defense of Pluralism and Equality*. Basic Books.

CHAPTER 3

The Missing Activities in the IPAT Equation

Mikko Jalas

1 Introduction

The IPAT equation has provided a ground for diverse debates on the environmental consequences of human activity.[1] According to the equation, environmental Impacts are produced by the factors of Population, average Affluence or economic turnover, and Technology with varying ratios of environmental efficiency. Yet, it is in many ways too abstract and general to provide useful insight into how we might react to negative environmental impacts (I). It is the technology factor that has received the most attention, including distinctions such as separating the factor for the energy intensity of production from a factor of decarbonising the energy supply. More detail is, however, not likely to solve the IPAT dilemma that human effort in technology development generates economic growth and population growth and thus undermines the efforts to reduce overall environmental burdens. The aim of this chapter is to consider modifications to the equation in order to support critical perspectives on sufficiency.

The equation has rendered a pursuit for, and critique of, technology development as a means to afford more wealth for more people. The view that P and A depend on T has been used both to issue techno-optimistic views on eco-modernism but equally to refute the belief that technological development could solve the crises.[1] Underlying this debate is the question of sufficiency and enoughness: how to settle with the wealth we have, how to check relentless economic growth and how to conceptually understand the key notion of enoughness.[2] While growth has obviously occurred both in the increasing size of human populations and in average levels of monetary and material consumption per capita, I will in the following focus on the latter. Affluence and Technology and their interconnected development stand for the qualitative changes in how humans have developed an exploitative and destructive relation to the biosphere. It also follows that if the aim is to alter

1 M. Chertow, "The IPAT equation and its variants," *Journal of Industrial Ecology* 4, no. 4 (2000).

2 J. Jungell-Michelsson and P. Heikkurinen, "Sufficiency: A systematic literature review," *Ecological Economics*, 195 (2022).

 | DOI:10.1163/9789004746404_004

such a relationship and work towards the peaceful co-existence of humans with other species, we need new and qualitatively different understandings of how to think about and organise human-nature relations. This requires the rethinking of both Affluence and Technology.

In this chapter, my aim is to rethink Affluence and Technology by relating them to time. In particular, I proceed to represent Affluence as *wealth-in-time* and *time affluence* measured as hours outside of paid work, and Technology as *resource intensity of time use*.[3]

Time affluence is related to a discussion on how the reduction of working hours is a necessary part of reducing the environmental impacts of human activity in absolute terms. It draws from the observations that the work time reduction, while rapid in the 20th century up to 1970, has stalled and that societies have opted for high levels of material wealth and long hours of work. This is a result of particular institutional and political arrangements, which do not necessarily correspond to human flourishing and happiness.[4]

Reframing technology as resource intensity of time use has perhaps even more radical consequences. It argues for a view that the major occupation for humans is simply to live a satisfactory, meaningful, decent life through a sequence of material and bodily entanglements. Put bluntly, humans "make" time and act out everyday practices by using and consuming products and interacting with their environments. A technology "improvement," when thinking in terms, refers not to increased output of production but to reduced resource use per unit of time. Consuming slower, and enjoying and appreciating the material engagements, suggest a new ethic of humble, sensory co-existence through which human orientation is not to manage labour and exercise self-discipline for maximised production to serve material wealth and future personal utility but rather in the here-and-now localised sensibilities and cross-overs between dichotomies such as mind/body and human/nature.

While this line of criticism is well established in environmental thought coming from diverse sources such as degrowth literature, utopian thinking of post-scarcity and environmental humanities, it has not been connected to the IPAT equation. Indeed, my interpretation is that the IPAT equation has mostly served techno-optimist views much in the way it was originally presented by

3 See: L. Reich, "Time and wealth," *Time & Society* 10, no. 2–3 (2001); S. Geiger et al., "Time wealth," *Current Research in Ecological and Social Psychology* 2 (2021); M. Jalas, "A time use perspective," *Ecological Economics* 41, no. 1 (2002); M. Jalas, "The everyday life context of increasing energy demands," *Journal of Industrial Ecology* 9, no. 1–2 (2005); M. Jalas and J. Juntunen, "Energy intensive lifestyles," *Ecological Economics* 113 (2015).

4 J. Schor, *The Overworked American* (Basic Books, 1991); G. Cross, *Time and Money* (Taylor & Francis, 1993).

Barry Commoner.[5] These premises stand in the way of making it a more critical tool for the debate and require tweaking, gendering, queering, or other forms of underground conceptual work. In the following, my aim is to place a de-centred, corporeal human subject into the IPAT equation by adding a factor of activities in the equation and suggesting a less instrumentalist view of technology. I proceed to first consider some conceptual resources for this effort through the notion of embodied experience and then follow up on the consequences for both Affluence and Technology.

2 Time as an Embodied Experience

My central argument is that economic theory and practice, in the form of, e.g., marketing and the development of new products and services, underlies the widespread and apparently omnipresent, relentless need for more stuff and as-if more content for life, the dissatisfaction with what one has, and ultimately the disability to engage with and enjoy life as it unfolds. The argument presented here aligns particularly with the ideas that the multi-sensorial human body and different skills of engaging and appreciating one's living environment give rise to alternative modes of being and offer oppositional stances toward the imperative of mass consumption, through-away society and the disengagement and apathy of contemporary consumption.[6]

Further lineage for my argument comes from the calls for de-centring the human subject and in more broad terms, pragmatism. Accordingly, human behaviour is to be understood not as instrumental rationality, self-interested utility maximation over time but through an embeddedness in the unfolding of everyday. Rationalities are fragmented as humans are thrown into different practical settings and perform practices as routinised behaviour. Instead of mastering and managing their environments, humans co-exist in and through a complex network of interactions. Ultimately, pragmatism portrays a humble view of the human being and offers an alternative ethico-political configuration than the one underlying seemingly de-political techno-rationalities.[7] It seems that the ideal modern rational subject has little alternative but to exploit

5 See: M. Schmelzer et al., *The Future is Degrowth* (Verso, 2022); A. Dobson, "Critical Theory in Green Politics," in A. Dobson and P. Lucardie, eds., *The Politics of Nature* (Routledge, 1993); K. Trebeck and J. Williams, *The Economics of Arrival* (Policy Press, 2019); T. Morton, *Being Ecological* (Pelican, 2018). On Commoner, see Chertow (2000).

6 K. Soper, "Re-thinking the Good Life," *Journal of Consumer Culture* 7, no. 2 (2007).

7 J. Dewey, *Art as Experience* (Penguin, 2005); N. Gross, "Pragmatism and the study of large-scale social phenomena," *Theory and Society* 47 (2018).

their environments, think about and optimise for their futures, and consequently pursue growth up to and beyond the biophysical limits. De-centred subjects meshed with shifting rationalities and corporeal experience of one's own existence hence offer a strong hold for radical environmental critiques.

These premises are not necessary parts of discussing time affluence. Often the debates on time affluence include the strive for better management of time, plannability, and control over time. These are issues that people report suffering from, but issues that do not deliver in terms of developing alternative and qualitative different modes of being for the prevailing instrumental rationality and human-centred exploitation. Similar, diluted versions of the critical thoughts around time and human experience can also be found in household economics, which allows for the fact that people have corporeal experiences and are not indifferent about them and consequently may reap "process-benefits" (or suffer) from activities, but in the end, does not allow these concepts to overthrow the assumption of an instrumentally rational human mind.[8]

Process benefits of human action and time politics, which call for more autonomy, self-control, and flexibility, appear to follow a pattern that Morton (2018) recognises concerning the treatment of aesthetics by Kant. Accordingly, authoritative theoretical treatments of human rationality recognise slippage and grey zones in which the human mind might be enchanted and mesmerized but theorize these in ways that rather purify the rational human subject rather than allow for critical de-centring of human subjects. Giving the embodied human experience a foreground over the rational, withdrawn human mind is a task that needs explicit attention. I will in the following attempt to pursue and locate these aims in an alteration of the IPAT equation to include human activity.

3 Redefining Affluence (A) as Time Affluence

The IPAT equation uses the term A (Affluence) to describe the level of well-being of human societies. It indicates the wealth drawn from economic activity and at the disposal of humans. Frequently, IPAT discussions have used Gross Domestic Product (GDP) to indicate the level of wealth. While GDP is a flow measure of added value rather than a stock measure of accumulated wealth, it does indicate the output of goods that goes into maintaining and reproducing the accumulated stocks of wealth.

8 G. Dow and F. Juster, "Goods, Time, and Well-Being," in F. Juster et al. eds., *Time, Goods, and Well-Being* (University of Michigan Press, 1985).

From the point of view of this paper, the GDP as a measure of wealth nevertheless has several downsides. Firstly, it is poor at registering affluence as a lived experience. There is a long discussion on how the focus on maximizing economic output (GDP) poorly correlates with subjective wellbeing.[9] Space does not allow to present the main claims and criticism thereof, but it suffices to note that the IPAT equation is subject to the wide range of criticism that relates to using economic output as the proxy of human wellbeing. The second deficit of GDP as a measure of affluence is that GDP and technological development are not independent of each other. Technology development supports the increase of economic output, and hence, long-term trends A and T are interlienked. Hence, increases in GDP easily absorb any improvements in technology, and while technology development may result in a decoupling of economic out and resource use, there is seldom an absolute reduction in the per capita environmental impacts, let alone signs of saturation of economic growth and stable economies.

Affluence can be accounted for in a non-monetized way. One of the critical strands of sustainability and sustainable consumption literature has focused on wealth-in-time and time affluence.[10] Time affluence provides another formulation of affluence, towards which the benefits of technology development could be directed. Stating the same with more insight into the biophysical limits of the planet, we cannot afford productivity increase and technology development unless we are able to direct it away from growth in economic output towards the increase of time affluence.[11] Historically, technology development has provided a time dividend and enabled the reduction of human toil and working hours. Yet, this development has been far from straightforward, but rather a political struggle since the birth of the labour movement in the 18th and 19th centuries and depended on the power and direction of this movement. Cross (2003), in fact, suggests that the recession in the 1930's and the second world war were culminating moments for this struggle to turn towards increased material output and material wealth over to reduced working hours. Since the 1970's this imperative seems to have got an increasing grip: the decline in working hours and the increase of time affluence has halted and even been reversed.[12] Even if more recent policy suggestions for work time

9 R. Layard, *Happiness* (Penguin, 2005).
10 See Reisch (2001) and Geiger et al. (2021).
11 C. Sanne, "Willing consumers—or locked-in?" *Ecological Economics* 42, no. 1–2 (2002).
12 https://stats.oecd.org/Index.aspx?DataSetCode=ANHRS

reduction have emerged from environmental politics, they remain absent or marginal in established policy processes such as the IPCC.[13]

Unlike monetary affluence, time affluence has a terminal value. In a country such as Finland, the average per capita working time is 2.8 hours per day when including everyone except children below ten years. The past reductions in the average working hours reflect different welfare policies such as the general provision of education and pension schemes and apparently more neutral developments such as the ageing of the population and increasing unemployment. It also holds that private investment in productivity growth across sectors creates options for worktime reduction. In these terms, Finns, in general, appear to be relatively time-rich, and there is only little scope for an advance towards time affluence. This does not, however, undo the importance of renouncing full employment in the formal economy as a social goal and the need for work-sharing as part of an agenda for more sustainable human societies. Reformulating the affluence factor A of the IPAT equation as time affluence emphasises this quality twist in thinking about sustainability.

4 Redefining Technology (T) as the Resource Intensity of Everyday Activity

In a formal equation such as IPAT, a change in one factor requires changes in others as well: reformulating Affluence as time affluence, requires the Technology T also to be changed. In the following, I pursue such a change by introducing factor C into the equation to describe how much consumption items go into particular activities. Combining C with T, and hence working with a new IPACT equation, provides a view of consumption "technology" as an account for how much resources are used per hour of human activity.

In this effort, I draw on a series of my previous works.[14] This chain of work was first and foremost initiated by my concern that eco-efficiency innovations such as fuel-efficiency of cars make room for more consumption and give rise to rebound effects, and that the technological narrative of efficiency innovations seemed not to reflect everyday realities or deliver in terms of absolute reductions of environmental harm.[15] In particular, the claims that "consumers

13 M. Pullinger, "Working time reduction policy," *Ecological Economics* 103 (2014); M. Antal et al., "Is working less really good for the environment?" *Environmental Research Letters* 16, no. 1 (2020).

14 Jalas (2002); Jalas (2005); Jalas (2015).

15 S. Sorrell and J. Dimitropoulos, "The rebound effect," *Ecological Economics* 65, no. 3 (2008).

do not need products, but only the service they yield" and that products could and should be thought of as "service-producing-machines" were, for me, signs of an inbuilt dilemma in the environmental debate.[16] While seemingly arguing for non-materialistic stances, these claim simultaneously portray rationalities that are not embedded in everyday existence but extend towards the future and assume ultimate, final needs that consumption and everyday practice is supposed to serve and be subordinate to. In doing so, these claims pre-empt meanings of everyday conduct and undermine the alternative, modest and humble ethical-political agendas that build on de-centring the human subject.

How, then, to avoid this trap and represent everyday practice as a potentially meaningful intrinsically motivated bundle of activities? First, the suggestion was rather simple: Rather than thinking about how to lower environmental impacts per unit of service, such as passenger kilometres, or unit of economic output, one should question and account for how much resources a single person needs per unit of time.

Resource use can, of course, be accounted for as annual averages, as, e.g., per capita consumption of energy services over a unit of time. Yet, gross figures of this kind would have little reference to everyday life and offer little insight into policy. Even when operationalised, such as 2000W society,[17] signalling an ambitious target for energy consumption per time, this continues to lack policy leverage. Ideally, one should be able to distinguish differences in the energy intensities of everyday life activity. This would be a measure of how much energy is needed or how much emissions are caused per unit of time in a given activity. It would formalise the intuitive truth that activities such as car driving require far more resources than those of, e.g., reading, craft and home entertainment, and that we can be quicker or slower in causing environmental harm. It could also give substantive and contextual grounds for practising sufficiency and strong sustainability, which would lead to an absolute reduction of the negative environmental consequences of human action.

5 IPACT: Formal Representation and Methodological Hurdles

In a reformulated IPAT equation, the term Affluence (A) is converted to represent time affluence as activity patterns outside of the hours of paid work. Factor C refers to consumption as how much goods and services are needed for

16 E. Weizsacker, et al., *Factor Four* (Club of Rome, 1998); M. Jalas, "Busy, wise, and idle time" (Helsinki School of Economics (2006).

17 https://www.2000-watt-society.org

a particular activity, and T remains to describe the emissions or harm caused by productive activities in the formal economy.

Impact (I) is thus a product of the following four factors:

- P – Population
- A – The sum of activities that make up the time of outside paid work
- C – The goods and services that are needed for the activities A
- T – Technology of production of consumption items C as the environmental harm created in the production of economic output

The methodological challenges for presenting human-induced environmental harm through a time-use lens are not trivial, but the method has been further formalised.[18] I will present a short summary of challenges relating to data and methods.

Time-use survey data exists to approximate the time used in different everyday engagements and provide the data for the Activity factor (A) with various levels of aggregation.[19] No depth of empirical detail will answer the question of how many activity categories one should use, or to say the least, this decision has an ethico-political dimension in regard to the aim to work towards de-centring the human subject. The more activities, the more overlaid but different and fragmented rationalities everyday life is suggested to have. Inserting a large number of activities leads to a very postmodern view of life as abundance: reading for the sake of reading, driving for the sake of driving, eating for the sake of eating, and painting the house or vacuum-cleaning for the sake of these very activities. Life is rendered thus not only affluent but also pointless and lacking higher causes and long-term projects. On the other hand, a low number of activities stands for subordinating activities for a higher goal, prescribing instrumental rationality as a prevailing orientation and valorising the clear and detached human mind and reasoning instead of the messy affairs of bodily engagements. As argued above, sufficiency debates will benefit from the former choice.

The consumption factor (C) contains another set of open questions. It requires the allocation of consumption expenditures to particular activities. Car driving requires at least the fuel, the car, the maintenance services, the insurance, and the road infrastructure. Reading requires a book, a chair, and a share of domestic energy use. Quite obviously, capital costs of car ownership, car maintenance, fuel and insurance would fall on the activity of driving. Less uncertain is how to allocate items such as all-purpose clothing or domestic heating that can be thought of as infrastructure or distant resources for other

18 J. Minx and G. Baiocchi, *Time Use and Sustainability* (Springer, Dordrecht, 2009).

19 https://ec.europa.eu/eurostat/web/time-use-surveys.

activities. In addition to drawing on distant resources, activities also interlink. Reading requires at least activities related to food and eating, and eating, in turn, for many, may require car use. Despite such interlinkages, it is obvious that people may end up spending a lot of time driving and little reading, or vice versa, with obvious resource implications.

The technology factor (T) remains essentially the same as in the traditional IPAT equation. It seeks to account for the environmental harm caused by producing goods and services in the economy. Empirically, techniques such as environmentally-extended input-output tables of national accounting (E-IO) leverage help. They have been used to capture and account for the various economic activities and related emissions that go into producing items for final consumption expenditure. Such a method, while by necessity an aggregate, summons up distributed harm into indices such as the carbon footprint per economic output in different sectors. This solves, in an acceptable way, the question of accounting for how global value chains connect to purchases of market commodities.

The IPACT equation developed in this chapter offers an important new view of how resource consumption happens as an everyday phenomenon. Factor C points out how households and individual people absorb and digest the output of the formal economy. When the consumption factor C is combined with production technology T, CT describes the resource- or harm-intensity of everyday activities, which in more lay terms accounts for how quick and effective humans are to induce environmental harm through consumption activities.

6 Discussion

Sufficiency, understood as enoughness, requires conceptual thinking on how societies and economies might mature and transcend the imperative of continuous economic growth. This will involve parallel consideration of both the material, substantial ecological crises in which humanity is far from being affluent or beyond scarcity, but also utopian thinking about abundance and post-scarcity, which turns the view away from survival towards more pacified relations amongst humans and between humans and other species.

This chapter has reformulated the IPAT equation in order to make room for post-scarcity and affluence, and a de-centred human subject to imagine and model human existence that is not based on survival and maximising one's own future utility. The modification proceeded on two interrelated tracks: Firstly, it provided better descriptive tools to understand and account for affluence (A) and rendered human existence in forms that allow for sufficiency, a slower pace of life, and less acquisitional modes of being. In essence, the proposed modification uses time-use survey data and substitutes "time affluence" for the affluence (A). Affluence is simply the sum of hours outside of paid labour.

The second track has extended the idea of technology to include both the production technology as in the traditional IPAT equation, but also consumption "technology" as in what products are needed in particular activities. Combining this idea of consumption with production technology yields a view of "resource use or caused harm per hour of human activity." This renders many contemporary consumption phenomena as technological regress and provides uncanny alternatives: all the way from the multiplication of special-purpose sports shoes to the speed-up of rail service, these changes increase the harm- (and money-) intensity of human activities of sport and travel, even if they are produced with efficient production technology. A slower pace of life, lower speed limits, pausing or sharing resources through, e.g., ride-sharing, "make," e.g., more "hours of driving" with same resources. When including a consumption view, this is technological progress, as resources per hour decline.

However, the potential effects of this tweak extend further. Whenever activities are introduced in this way, the equation claims them to be meaningful in their own right. Put simply, mundane activities such as cooking, cleaning, maintenance, and also shopping become the high moments of everyday existence and the dichotomy between domestic work and leisure is abolished. In conceptual terms, the tweak prompts the ideas of plenitude and abundance, and "arrival" as opposed to scarcity, the need to economise resources, and be productive. It reflects hope and dire need to acknowledge that humans, at some point and somehow, might have enough.

Whilst the traditional version of IPAT leaves these questions untouched and does not inquire into the purpose, aims or framings of consumption, the reframing of the equation as IPACT does this in a radical way. It interrogates us in the question of terminal values and meaningfulness in everyday life. Whenever, for example, accounting for resources and time that go into cooking, cooking is represented and framed as a terminal value. Hence, affluence and plenitude and "arrival" are not only the infrequent and rare accomplishments or the highlights of leisure, but omnipresent companions of human experience in a much more punctuated mode.

This tweak has given rise to a narrow stream of empirical research seeking to deliver descriptive statistics on materials, energy, and carbon intensity of human activity.[20] However, the radical potential of the time use perspective

20 A. Druckman et al., "Time, gender, and carbon," *Ecological Economics* 84 (2012); J. Heinonen, "Situated lifestyles," *Environmental Research Letters* 8, no. 3 (2013). Madsen, S. T., & Weidema, B. P. (2023). Input–output modelling for household activity-level environmental footprints: a systematic literature review. Environmental Research Letters, 18, no. 4, 043003.

has often not been appreciated but rather seems to disappear in empirical detail. The sufficiency politics involved in this approach hence have rarely surfaced in this literature, even if they are pronounced as the calls for shorter working hours as part of wider sufficiency debates.

Previous work on the methods has mainly emerged from household economics. For example, Gronau and Hamermesh investigate the "goods intensity" of non-market activities, i.e. expenditures per unit of time.[21] While there is an unarticulated radical point in this analysis, i.e. that there is no hierarchy of and no intermediate outputs in non-market activity and activities seem to be self-serving and not instrumental towards other activities, household economics present a tidy and muted version of the critique advanced in this chapter. Either the interest in time-use is driven by an intention to represent everyday activities as production that combines market goods and consumers' time to produce specific non-market goods, or they include a notion of "process benefits" or the possibility of qualitative experience in everyday life as a sidenote to the economic theory of human behaviour. It remains that household economics fails to carry this way of thinking to the available logical conclusion of displacing economic rationality with a perspective of dwelling in and co-existing with multiple overlapping rationalities.

The reasons for such a failure of household economics may be many. If the reified belief in human rationality and exceptional status amongst species is as strong as Morton (2018) claims, with roots not only in enlightenment philosophy but also in monotheist religions and agricultural practice, thinking human existence in terms of de-centred, aesthetically and emotionally enchanted subjects is deeply disturbing. There are also good reasons to think that such a position will be increasingly difficult to adopt in the middle of environmental turmoil, which seemingly calls for more cool-headed management of unfolding crises.

These efforts can also be understood in terms of the prevailing ideology of busyness. The imperative to economize one's time use is deeply rooted in contemporary societies. To oppose it, to regard time as abundant instead of scarce, is really to think the unthinkable, as Foucault urges us to do (cited in Darier [1999]) or turning things on their head with the help of the psycho-analysis of Lacan. Time is a scarce resource for life projects; our minds and cultural resources reassure us!

How might this reformulation and an IPACT equation help address the substantial concerns of environmental degradation, nature loss, global warming,

21 R. Gronau and D. Hamermesh, "Time vs. goods," *Review of Income and Wealth* 52, no. 1 (2006).

and overconsumption of resources? A number of points can be excavated, although their further elaboration and systematization is beyond the scope of this text. Firstly, if the rising level of affluence is considered a problem as in the traditional IPAT equation, does the same apply to time affluence. Production technology is pushing people out of work. Productivity increases through technology free humans from labour but simultaneously require intensive consumption during the hours of leisure. Increased time affluency makes the task of absorbing production output easier. Work life can also chance. Less pressured work in a less competitive economy, which would not be driven by the maximization of labour productivity and which would not yield more hours of leisure, could halt and provide a check on rampant resource extraction. Furthermore, if the productivity increases at work decline or halt, which could also be the case with the ending of the fossil era, shorter hours of work and pursuit for increases time affluence will be less needed in reducing overall economic through-put.

The reformulated view of technology as a combination of production technology T and consumption technology C also offers relevant policy insights. When considering, e.g., the objectives of cities to provide public space for local free time activities as opposed to long-distance travel, parallel interests are present. These policies can yield affordable alternatives pastime activities that cause fewer negative impacts per unit of time. Such strategies, on the other hand, are often compromised by a broad range of forces of acceleration. Fast rail and other forms of speeding up mobility are large public financial investments of fitting more consumption into the everyday, and, as the IPACT accounts, thereby regressive technology changes as more resources are needed by a unit of time.

The political tensions around busyness can perhaps be best appreciated by recognising the dual presence of the imperative to economise one's time and the empirical evidence that people often regard their personal wellbeing to depend on the opportunities to exchange material wealth for more free time. Time affluence is not unthinkable. It is rather a sort of pragmatic reasoning of the experiencing human body and a critical reflection by individuals of the societies that are plagued by over-production and intensive marketing and sales promotion. It is only in the realm of economic theory that the pursuit of time affluence is rendered as a short-sighted, myopic pursuit of present pleasure at the expense of the (growth of) future consumption opportunities. It is obvious that treating time as a form of final utility will radically change the way we think about technology and human subjects. To be sure, human existence is marked by intentionality, abilities to think about the future, make tools, develop technology, and solve problems. Nevertheless, at the same time,

time affluence and the appreciation of our corporeal existence are no less trivial parts of human existence.

7 Conclusion

Formal representations such as the IPAT equation are tools for thinking that also forge particular rationalities, politics, and ethics. This chapter has sought to embed a respectful and modest ethical-political subject position in the simple IPAT equation and remodeled it into an IPACT equation by introducing time affluence as a form of affluence. In doing so, it has also led to the redefinition of the Technology factor of the IPAT equation in ways that endorse slower living, sharing and the possibilities of finding contentment and joy across a range of everyday activities. The ethos of the text is to tread lightly on the earth. Metaphorically and literally, sufficiency requires that we lead our life, leaving little traces as we go on with everyday conduct. It is to ask what the conditions, consequences and cruxes of our momentary existence are, day to day and hour to hour.

Acknowledgements

This work has been supported by the Strategic Research Council grant no 13352478 at the Academy of Finland.

References

Antal, M., Plank, B., Mokos, J., and Wiedenhofer, D. (2020). "Is working less really good for the environment? A systematic review of the empirical evidence for resource use, greenhouse gas emissions and the ecological footprint." *Environmental Research Letters* 16, no. 1, 013002.

Chertow, M. (2000). "The IPAT equation and its variants." *Journal of Industrial Ecology* 4, no. 4: 13–29.

Cross, G. (1993). *Time and Money: The Making of Consumer Culture*. Taylor & Francis.

Darier, E. (1999). "Foucault against Environmental Ethics," in *Discourses of the Environment*, edited by E. Darier. Blackwell.

Dewey, J. (2005). *Art as Experience*. Penguin.

Dobson, A. (1993). "Critical Theory and Green Politics," in *The Politics of Nature*, edited by A. Dobson and P. Lucardie. Routledge.

Dow, G. and Juster, F. (1985). "Goods, Time, and Well-Being: The Joint Dependence Problem," in *Time, Goods, and Well-Being*, edited by F. Juster and F. Stafford. University of Michigan Press.

Druckman, A., Buck, I., Hayward, B., and Jackson, T. (2012). "Time, gender and carbon: a study of the carbon implications of British adults' use of time." *Ecological Economics* 84: 153–163.

Geiger, S. et al. (2021). "Time Wealth: Measurement, Drivers and Consequences." *Current Research in Ecological and Social Psychology* 2, 100015.

Gronau, R. and Hamermesh, D. (2006). "Time vs. Goods: The value of measuring household production technologies." *Review of Income and Wealth* 52, no. 1: 1–16.

Gross, N. (2018). "Pragmatism and the Study of Large-scale Social Phenomena." *Theory and Society* 47: 87–111.

Heinonen, J. et al. (2013). "Situated Lifestyles: II. The impacts of urban density, housing type and motorization on the greenhouse gas emissions of the middle-income consumers in Finland." *Environmental Research Letters* 8, no. 3, 035050.

Jalas, M. (2002). "A time use perspective on the materials intensity of consumption." *Ecological Economics* 41, no. 1: 109–123.

Jalas, M. (2005). "The everyday life context of increasing energy demands: Time use survey data in a decomposition analysis." *Journal of Industrial Ecology* 9, no. 1–2: 129–145.

Jalas, M. (2006). "Busy, Wise and Idle Time: A study of the temporalities of consumption in the environmental debate." Helsinki School of Economics.

Jalas, M. and Juntunen, J. (2015). "Energy intensive lifestyles: Time use, the activity patterns of consumers, and related energy demands in Finland." *Ecological Economics* 113: 51–59.

Jungell-Michelsson, J. and Heikkurinen, P. (2022). "Sufficiency: A systematic literature review." *Ecological Economics* 195, 107380.

Layard, R. (2005). *Happiness: Lessons from a New Science*. Penguin.

Madsen, S. T., & Weidema, B. P. (2023). Input–output modelling for household activity-level environmental footprints: a systematic literature review. Environmental Research Letters, 18, no. 4, 043003.

Minx, J. and Baiocchi, G. (2009). "Time use and sustainability: An input-output approach in mixed units," in *Handbook of Input-Output Economics in Industrial Ecology*. Springer, Dordrecht.

Morton, T. (2018). *Being Ecological*. Pelican.

Pullinger, M. (2014). "Working time reduction policy in a sustainable economy: Criteria and options for its design." *Ecological Economics* 103: 11–19

Reisch, L. (2001). "Time and Wealth." *Time & Society* 10, no. 2–3: 367–385.

Sanne, C. (2002). "Willing consumers—or locked-in? Policies for a sustainable consumption." *Ecological Economics* 42, no. 1–2: 273–287.

Schmelzer, M., Vetter, A., and Vansintjan, A. (2022). *The Future is Degrowth: A Guide to a World Beyond Capitalism.* Verso Books.

Schor, J. (1991). *The Overworked American: The Unexpected Decline of Leisure.* Basic Books.

Soper, K. (2007). "Re-thinking the Good Life: The citizenship dimension of consumer disaffection with consumerism." *Journal of Consumer Culture* 7, no. 2: 205–229.

Sorrell, S. and Dimitropoulos, J. (2008). "The rebound effect: Microeconomic definitions, limitations and extensions." *Ecological Economics* 65, no. 3: 636–649.

Trebeck, K. and Williams, J. (2019). *The Economics of Arrival: Ideas for a Grown-up Economy.* Policy Press.

Weizsäcker, E., Lovins, A. B., and Lovins, L. H. (1998). *Factor Four: Doubling Wealth – Halving Resource Use: The New Report to the Club of Rome.* Earthscan.

PART 2

Sufficiency in Affluence

CHAPTER 4

From Overconsumption to Respecting Planetary Boundaries

Tina Nyfors and Lassi Linnanen

1 Introduction

The state of the environment is deteriorating worldwide, at an unparalleled scale, threatening the life-sustaining conditions of the biosphere on which humans are completely dependent. Climate change, biodiversity loss, and pollution and waste are three interconnected planetary emergencies caused by human activities. Rising inequality in multiple dimensions affecting human well-being, including socioeconomic inequality, represents a fourth challenge with cross-cutting impacts.

The magnitude of the human impact on the Earth has given rise to the notion that the planet has entered a new geological epoch: the Anthropocene. To date, seven out of nine planetary boundaries have been crossed.[1] According to the World Meteorological Organization (2025), the greenhouse gas (GHG) emissions caused by human activities are higher than in the past 800 000 years and are showing a rising trend. The effects of climate change include extreme weather events like droughts, floods, and storms becoming increasingly visible around the world, exacerbating pressure on freshwater resources and affecting, for example, food security.

Concurrently, biodiversity—the diversity within species, between species, and of ecosystems—is declining faster than at any time in human history. From a biodiversity perspective, human impact on the environment has increased particularly since the 1970s, leading to what has been called a sixth mass extinction. According to IPBES (2019), one fourth of all animal and plant groups assessed were found to be threatened by human actions, and at the current trajectory, an estimated one million species face extinction, many of

1 K. Richardson, W. Steffen, W. Lucht et al., "Earth beyond Six of Nine Planetary Boundaries," *Science Advances* 9, no. 37 (2023); Planetary Boundaries Science, "Planetary Health Check 2025" (Potsdam Institute for Climate Impact Research, 2025). Planetary Health Check 2025: https://www.stockholmresilience.org/news--events/general-news/2025-09-24-seven-of-nine-planetary-boundaries-now-breached.html, "Seven of nine planetary boundaries now breached".

 | DOI:10.1163/9789004746404_005

these within decades. The harmful impacts of human land use, resource overexploitation, and pollution are further exacerbated by climate change.

A symbolic threshold regarding human impact on Earth was crossed when the human-made mass surpassed the weight of the overall living biomass in the world around 2020; similarly, the mass of produced plastic was found to be greater than the mass of all animals living on land or in water.[2]

During the past 50 years, the population has doubled, as has the average affluence of the population, while technology enables unprecedented human impact on Earth. Of the three interlinked factors in the IPAT equation, we find that sufficiency most easily lends itself to discussions on affluence (see also Hayden 2020) and, connected to this, technology. The current lifestyle of the rich underscores the need of reaching a good life within planetary boundaries particularly among the affluent.

Households stand for approximately two thirds of the climate impact globally according to consumption-based accounting.[3] Wealthy individuals contribute disproportionately to higher GHG emissions, which illuminates that it is not meaningful to refer to 'humanity' as a unified group. High emitters live in rich as well as in poor countries. The richest 10% of household emitters globally are responsible for half of all carbon dioxide (CO_2) emissions, whereas the richest 1% alone emits as much (16%) as the poorest 66% of the global population.[4] According to some estimates, an even smaller share of the richest (0.54%) stand for 14% of the global GHG emissions.[5]

The Paris Agreement (United Nations 2015) outlines a limit for climate change as staying well below 2 degrees of global warming, and the Convention on Biological Diversity (United Nations 2022) outlines a target of protecting 30% of the land and sea areas by 2030. To reach these targets, international reports repeatedly call for fundamental, rapid change. Currently, there is no country where citizens' basic needs are met at a level of resource use that would be globally sustainable; however, according to researchers, strategies "with a focus on sufficiency and equity, have the potential to move nations towards sustainability, but the challenge remains substantial."[6]

The pressing environmental problems combined with inequality raise sufficiency-oriented questions about "what is enough," "what is a good life", and

2 E. Elhacham, L. Ben-Uri, J. Grozovski et al., "Global Human-Made Mass Exceeds All Living Biomass," *Nature* 588, no. 7838 (2020).

3 UNEP, "Emissions Gap Report 2020" (United Nations Environment Programme 2020).

4 A. Khalfan, A. Nilsson Lewis, C. Aguilar et al., "*Climate Equality: A Planet for the 99%* (Oxfam 2023).

5 I. M. Otto, K. M. Kim, N. Dubrovsky, W. Lucht, "Shift the Focus from the Super-Poor to the Super-Rich," *Nature Climate Change* 9, no. 2 (2019).

6 D. W. O'Neill, A. L. Fanning, W. F. Lamb, and J. K. Steinberger. "A Good Life for All within Planetary Boundaries," *Nature Sustainability* 1, no. 2 (2018), 88.

questions regarding the difference between needs and 'wants.' Sufficiency has brought up aspects related to both overconsumption and deprivation, both regarding people and countries. However, the sufficiency concept has been criticized for lacking ways of operationalization.[7]

This chapter aims to address this gap. We develop seven analytical steps for establishing sufficiency-based proposals by linking the concepts of sufficiency, environmental footprints, and planetary boundaries. The footprints are a way of measuring human pressure on the environment, planetary boundaries aim to define thresholds and a safe operating space for humanity, whereas sufficiency is a strategy for reducing human pressure on the environment, along the lines of "a good life for all within planetary boundaries"[8] (O'Neill et al. 2018). The carbon footprint is chosen as an example to inform of the sufficiency approach, while the water and material footprints are used to illustrate the implications of applying sufficiency proposals to other footprints.

To address the 'affluence' factor of the IPAT equation, we draw particular attention to the footprints of various income classes and to the most carbon-intensive options, i.e. footprint 'hotspots.' We propose that operationalizing sufficiency by using an income class perspective, firstly, helps crystallize various sufficiency-based proposals for different income classes, and secondly, adds a more positive dimension to sufficiency by making visible that a 'sufficiency class' *already exists*, pointing to new role models. In the transformation from overshoot to sufficiency, consumption and production represent two sides of the same coin, and sufficiency needs to be applied to both. This chapter focuses on the consumption perspective.

Based on calculations in previous literature, we develop sufficiency-based proposals, which can be beneficial for understanding the *scale* of cultural transformation needed. The sufficiency-based proposals for income classes are, in our understanding, well compatible with, and can support, wider understandings of sufficiency. As our purpose is to illustrate the operationalization of sufficiency, we do not outline the deep societal and cultural transformation needed. Furthermore, our analysis regarding footprint hotspots and their relation to income classes is non-exhaustive and illustrative. The focus is largely anthropocentric—as the environmental challenges originate from human activities, our search for solutions concentrates on humans. We recognize that the concepts for measuring human pressure and environmental thresholds have limitations, nevertheless the message is overwhelmingly clear: human impact on the environment needs to be curbed to preserve a livable planet.

7 J. Jungell-Michelsson and P. Heikkurinen, "Sufficiency," *Ecological Economics* 195 (2022).

8 D.W. O'Neill, "A good life for all within planetary boundaries," *Nature Sustainability*, 1, no. 2 (2018).

2 Footprints, Planetary Boundaries, and Sufficiency

2.1 *Environmental Footprints*

Measuring the human pressure on Earth has received increased interest, and environmental footprints are probably the most used tool.[9] Impact has been assessed at various scales, from global to regional and local levels, with a focus on nations, cities, individuals, organizations, or specific products.

Providing a clear footprint definition is challenging, as there are various understandings of the concept. The ecological footprint was the first footprint indicator, developed in the 1990s. Originally, footprint referred to the area of land needed to sustain an urban region, with the aim of providing an alternative to GDP. The ecological footprint has received widespread attention both in research and in public discussion, for example, through the 'Earth overshoot day.'

Over time, the footprint definition developed to refer to a more generalized indication of human pressure on the environment, as not all human pressures on the environment can be directly translated to land area. Many additional environmental footprint indicators have been developed. These include, for example, the carbon footprint (or climate, greenhouse gas footprint), which is the most well-known indicator, developed during the early 2000s, and the water footprint likewise proposed in the early 2000s, building on work on virtual water in the 1990s. In addition, there are nitrogen, phosphorous, material, biodiversity, particle matter, and energy footprints, to mention a few. Also, socioeconomic footprints have been developed, such as the employment, labor, inequality, and celestial footprints.

Differences between the footprints concern, for example, methodology, where two main approaches are employed—the original environmental footprint assessment (EFA), focusing primarily on national and global footprints, and the life cycle assessment (LCA) approaches that focus on products and organizations. Other differences include units of measurement, data availability, and whether focus is on resource use or emissions, or both. There are also overlaps between footprints. This has led some to call for unification and standardization of the footprints, whereas others suggest that attempts should not even be made to create a universal footprint definition but rather appreciate the diversity of the different footprints.

An emerging idea is combining multiple footprints into 'footprint families',[10] as no single indicator can capture the complexity of the human environmental

9 J. Matuštík and V. Kočí. "What Is a Footprint?," Journal of Cleaner Production 285 (2021).

10 D. Vanham, A. Leip, A. Galli et al. "Environmental Footprint Family to Address Local to Planetary Sustainability and Deliver on the SDGs," *Science of The Total Environment* 693 (2019).

impact. This is a way to reduce risks of trade-offs, where reducing one footprint increases the impact of others. Taking the example of wood, climate policies leading to increased bioenergy use may result in accelerated biodiversity loss.

2.2 *Planetary Boundaries*

Probably the most well-known attempt to quantify safe boundaries for human activities is the planetary boundaries concept, which, according to Rockström et al. (2009a: 472, see also 2009b), aims to define "the safe operating space for humanity with respect to the Earth system and are associated with the planet's biophysical subsystems or processes." The researchers further distinguish between thresholds, biophysical processes that if crossed, will "trigger non-linear, abrupt environmental change" and boundaries referring to values determined by humans, set at what is considered a 'safe' distance from biophysical thresholds or from a 'dangerous level' in cases where no clear thresholds exist.

Nine Earth-system processes have been identified by Rockström et al. along with thresholds that "if crossed, could generate unacceptable environmental change." The processes include climate change, biosphere integrity (originally referred to as biodiversity loss), freshwater change, biogeochemical flows, ocean acidification, novel entities, stratospheric ozone depletion, atmospheric aerosol loading, and land system change. The framework has been updated to include regional-level boundaries for five planetary boundaries, for example, freshwater use, change in land use, and biosphere integrity.

To date, seven boundaries are exceeded, including climate change, biosphere integrity, freshwater change, biogeochemical flows, novel entities, land system change, and most recently, ocean acidification.[11] The planetary boundary concept has been criticized for not addressing questions related to justice, and for lacking support in the Global South, for example.[12]

2.3 *Sufficiency*

Strategies to reduce environmental impact have commonly focused on increasing efficiency and technical solutions, meaning a reduction of emissions and resource use *per unit* of goods and services or in relation to economic output. Despite remarkable increases in efficiency, *total* global GHG emissions have repeatedly reached new record levels. Known as the "rebound effect," this is due to efficiency measures tending to lead not only to lower emission levels but also to lower prices, which in turn leads to increased consumption, and

11 Richardson et al. (2023); Planetary Boundaries Science (2025).

12 F. Biermann and R.E. Kim, "The Boundaries of the Planetary Boundary Framework," Annu. Rev. Environ. Resour., no. 45 (2020); A.S. Downing et al., "Learning from generations of sustainability concepts," Environ. Res. Lett., no. 15 (2020).

hence, *increased overall* emissions. Sufficiency represents an alternative and complementary strategy, focusing on *absolute* reduction.[13]

Central in sufficiency is a notion of 'enoughness' to remain within ecological boundaries.[14] Sufficiency gives rise to questions such as "what is a good life", while pointing to the need of distinguishing between needs and wants or desires.[15]

In this chapter, we align with the stream of sufficiency research approaching sufficiency as both an upper and lower boundary to consumption—enoughness as a maximum that should not be surpassed in terms of environmental impacts and as a minimum that everyone should be granted.[16] Similar ideas are developed in the 'environmental space,' 'consumption corridor,' and 'doughnut economy' concepts. In this chapter, particular attention is given to the upper boundary.

2.4 *Linking the Concepts*

There is an awakening discussion on linking environmental footprints with the idea of planetary boundaries, and, for example, Wu et al. (2021) call for "planetary boundary-based footprint family research." Combining the concepts is highly relevant from a policy-making perspective, as the planetary boundaries can be used to interpret footprint results.

Sufficiency outlines a vision of 'enough' within boundaries. Combined with the footprint and planetary boundaries concepts, it implies that some need to sharply reduce their pressure on the environment, whereas others need to increase theirs to fulfill basic needs. When linking the three concepts, it is possible to outline sufficiency-based proposals for various segments of the population, based on whether they are above, at, or below a sustainable level.

3 Operationalizing Sufficiency

Our suggestion for operationalizing sufficiency builds on Ulvila and Wilén's (2017) work on income classes, which builds on Chancel and Piketty (2015). Ulvila and Wilén develop their classification based on two criteria: fulfillment of basic needs and environmental impact. They note that both are linked to income, which is more tangible when considering solutions. They distinguish between four income classes linked to the scale of environmental damage:

---

13 B. Alcott, "The Sufficiency Strategy," *Ecological Economics* 64, no. 4 (2008).

14 T. Princen, *The logic of sufficiency* (2005); Jungell-Michelsson and Heikkurinen (2022).

15 See Alcott (2008).

16 See, e.g., L. Spengler, "Two Types of 'Enough'," *Environmental Politics* 25, no. 5 (2016).

over-consuming (richest 10% of the global population, causing 'very high' environmental damage), consuming (one fifth of the global population, causing 'significant' environmental damage), sustainable (more than one third, causing 'moderate' environmental damage), and struggling (one third, causing 'small' environmental damage). Basic needs are utilized to identify the thresholds between the struggling and sustainable classes. They do not estimate numeric footprints for the different classes.

For establishing sufficiency proposals, we develop the class perspective further. Firstly, the world population is divided into seven income classes: 1) Ultrarich overconsuming, 2) Overconsuming, 3) Consuming (in rich countries), 4) Consuming (in China, India, and other regions of the world) 5) Sufficiency 6) Struggling (in rich countries) and 7) Struggling (in China, India, and other regions of the world). Among the overconsuming, i.e. the richest 10%, we direct attention also to the ultrarich overconsuming class, which constitutes the richest 1% of the global population. Furthermore, we separate between the consuming class in rich countries and in China, India, and other regions of the world. For the purpose of this study, we have renamed the sustainable class and call it the 'sufficiency class.' Finally, we address the struggling class in poor and rich countries separately as the carbon footprints of the hundreds of millions poorest citizens are considerably lower than those of the poorest living in OECD countries.

Rich people live on all continents. Half of all GHG emissions are caused by the richest 10% of the people, of which individuals in the United States and Canada (hereafter referred to as North America, excluding Mexico) are responsible for 30%, followed by 15% by the European Union and the UK and 15% by China.[17] In particular, North America and the EU will be examined in more detail.

When discussing the richest 1% and richest 10% of the population, depending on data availability, we sometimes refer to numbers that do not exactly correspond to the global income segments of the population, e.g. the richest 10% or the richest 1% in the EU. Despite not corresponding to the global segments, we consider the numbers informative. We also acknowledge that not all individuals in an income class have the same environmental impact, hence, the numbers can be seen as indicative. Furthermore, lower-income regions are exemplified by China and India, chosen due to their large populations. It is worth noting that, as circumstances vary across regions, environmental impact corresponds to various income thresholds among countries in the same income class.

17 Khalfan et al. (2023) Methodology note, p. 10.

3.1 *Establishing Sufficiency-Based Proposals*

To establish sufficiency-based proposals we have followed seven analytical steps (see Figure 4.1). First, a specific planetary boundary is addressed (step 1) and translated to maximum per capita footprints (step 2). Then the correlation between income and footprint is established (step 3). To reduce the environmental footprint, the next step is to identify the most harmful sources, i.e. the 'hotspots' (step 4). After this, the hotspots are scrutinized for each income class, resulting in average 'hotspot footprints' for each class (step 5). Then a sufficiency perspective is applied by placing the hotspot footprints of the income classes in relation to planetary and social boundaries (step 6). The final step is establishing sufficiency proposals for each of the income classes, focusing on hotspots (step 7).

In the following, the steps for establishing sufficiency-based proposals will be outlined for the planetary boundary of climate change.

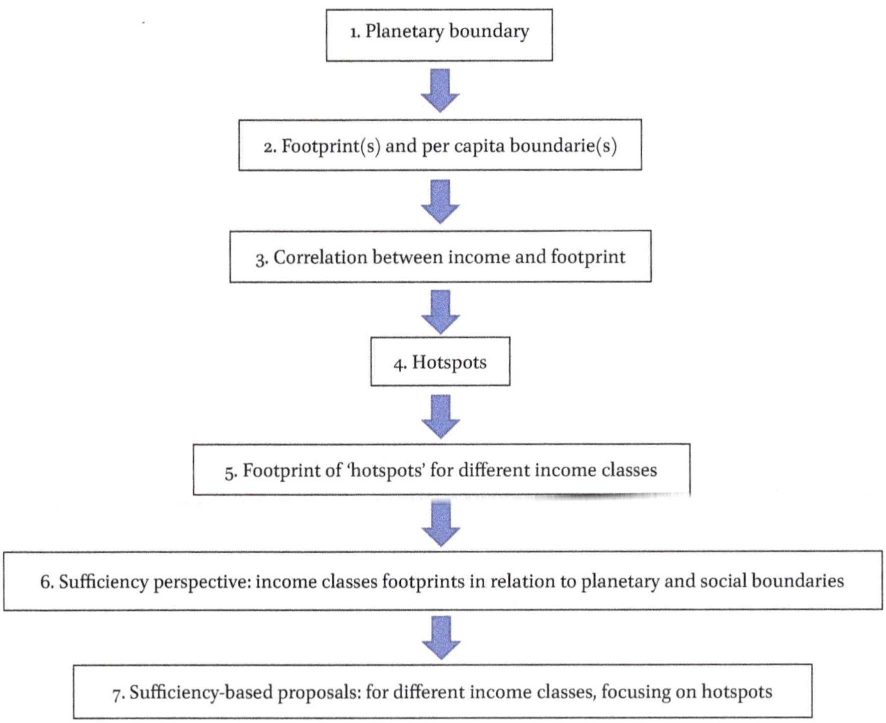

FIGURE 4.1 Analytical steps to establish sufficiency-based proposals for environmental footprints

3.1.1 From Planetary Boundary to Carbon Footprint and Per Capita Boundary (Steps 1 and 2)

The planetary boundary for climate change is estimated to be 1.5 or at maximum 2 degrees global warming. This planetary boundary corresponds to the carbon footprint, which can be defined as the "total amount of carbon dioxide emissions that is directly and indirectly caused by an activity or is accumulated over the life stages of a product."[18]

The climate change planetary boundary can be translated into a per capita carbon footprint boundary of around 2.5 tCO2e by 2030.[19] To reach the necessary reductions in GHG emissions, the per capita target footprint is reduced stepwise during the following two decades, to 1.4 tons per capita by 2040 and to 0.7 tons per capita by 2050. Other estimates provide numbers of similar magnitude.

3.1.2 Income Classes and Carbon Footprints (Step 3)

The correlation between income class and footprint is established in the following. The link between carbon footprint and income is well documented—as the Intergovernmental Panel on Climate Change (IPCC) notes, income is one of the most important predictors of carbon footprint size; higher income tends to correspond to larger footprints.[20] The carbon footprints and income thresholds for the different income classes are summarized in Table 4.1. An overview of the share of the global population belonging to each income class is given in Table 4.2. The table footnotes include the data sources of sub-section 3.1.2.

Related to table 4.2, in the consuming and struggling classes, a rough division is made between 'rich countries' and 'China, India, and other regions of the world.' The share of population in the income classes is indicative, building on Ulvila and Wilén (2017) and the numbers are approximate, to give a sense of magnitude. The global population is 8 billion; total population in the largest regions and countries include North America 370 million, EU 450 million, China 1410 million, and India 1410 million.

3.1.2.1 *The Ultrarich Overconsuming Class*

The ultrarich overconsuming class constitutes the richest 1% of the world population, with annual incomes above 220 000 Euros. The carbon footprints are estimated to be in the range of 200—320 tCO2e/capita per year, Chancel and Piketty (2015) argue that this "can be seen as a plausible value for the 1% richest individuals of this planet." Other studies using different methods suggest considerably lower carbon footprints for the ultrarich, between 50 and 65 tCO2e/capita per year. The 1% richest in the EU are estimated to have average carbon footprints of 55 tCO2e/capita per year.

18 T. Wiedmann and J. Minx, "A Definition of 'Carbon Footprint'', in C.C. Pertsova, ed., *Ecological Economics Research Trends* (Nova Science Publishers, 2008).

19 L. Akenji, M. Bengtsson, V. Toivio et al., *1.5-Degree Lifestyles* (Hot or Cool Institute, 2021).

20 M. Pathak, R. Slade, P.R. Shukla et al., "Technical Summary," in P. Shukla et al., eds. *Climate Change 2022* (Cambridge University Press, 2022). Chapter 2 p. 62.

TABLE 4.1 Income classes, carbon footprints, and income thresholds (the data sources are indicated in square brackets and listed in the footnote)[a]

	Ultrarich over-consuming (richest 1%)	Over-consuming (richest 10%)	Consuming (18%)		Sufficiency (40%)	Struggling (32%)	
			Rich countries (6%)	China, India, and other regions of the world (12%)		Rich countries (1%)	China, India, and other regions of the world (31%)
Carbon footprint tCO2e/ person per year	Non-country specific: 200–320 [2] 64.7 [5] US (richest 2%): 53.5 [1] EU (richest 1%): 55 [6]	Average: 18 [3] US: 50 [3] Canada: 30 [3] EU: 15–48 (average 23) [6]	Average: 7 [2]		–5 Basic needs met, not causing excessive CO_2 emissions	Small to moderate	Very small [4]
Income threshold EUR/cap/ year	220 000 [2]	North Am&EU: 14 000 China: 5700 India: 14 000 [4]	North Am&EU: 6000 [4]	India: 5600 China: 2700 [4]	700–6000 [4]	Deprivation of a decent life, may be due to homelessness, unemployment, inadequate access to healthcare	<700 [4]

[a] Data sources: [1] Ummel (2014); [2] Chancel and Piketty (2015); [3] Oxfam (2015); [4] Ulvila and Wilén (2017); [5] Otto et al. (2019); [6] Ivanova and Wood (2020)

TABLE 4.2 Number of people globally representing the different income classes (millions) (building on Ulvila and Wilén 2017)

	Ultrarich over-consuming (richest 1%)	Over-consuming (richest 10%)	Consuming (18%)		Sufficiency (40%)	Struggling (32%)	
			Rich countries (6%)	China, India, and other regions of the world (12%)		Rich countries (1%)	China, India, and other regions of the world (31%)
North America	35	220	110	–	20	20	–
EU	15	135	240	–	55	20	–
Other rich countries	12	50	70	–	30	20	–
China	8	150	-	445	705	–	125
India	1	30	-	30	760	–	600
Rest of the world	8	200	-	520	1630	–	1800
TOTAL	80	800	450	990	3200	60	2500

The largest share of ultra-rich individuals lives in North America, particularly the US, and in Luxemburg, Singapore, and Saudi Arabia. The carbon dioxide emissions by the richest 0.1% and 0.01% are discussed in more detail by Khalfan et al. (2023).

3.1.2.2 *The Overconsuming Class*

The overconsuming class is the richest 10% of the world population, living predominantly in North America and the EU, with incomes above 14 000 Euros. This class has average footprints around 20 tCO2e/capita per year. The variations in both income and footprints are huge, given the wide income range. The largest share of the richest 10% live in North America, where 60% of the population belongs to this class. This is followed by the EU with a third of the population belonging to this class. China has the third most individuals belonging to the overconsuming class, however, constituting a smaller share of the Chinese population. India is a telling example regarding income differences in various parts of the world. The income threshold for the overconsuming class in India is the same as in North America and the EU, but only around 2% of the population belong to this class, totaling around 30 million people. This illustrates how an income above 14 000 Euros may be much in Delhi but little in, say, Helsinki. While the overconsuming class is the dominant income class in Finland and second biggest in the EU, it is represented by only a fraction of the population in India.

Large regional differences occur in carbon footprints. North America tops the list, with the US standing out with an average carbon footprint of up to 50 tCO2e/capita per year. For the EU, 30% of the population belong to the global richest 10%.[21] While lacking data on this group as a whole, the richest 10% in the EU has an average carbon footprint around 20 tons, in line with the global average for the overconsuming class. Chinese constitute 20% of the richest tenth in the world, with a share of 15% of this income class' total CO_2 emissions.[22] For India, Oxfam (2015) estimates that the richest 10% have footprints corresponding to one quarter of the poorest half of the US population. Overall, 40% of the emissions are caused by North America and the EU, whereas China and India together stand for around a fifth.[23]

3.1.2.3 *The Consuming Class*

The consuming class constitutes almost a fifth of the global population. Incomes in the EU, North America, and other rich countries range from 6000

21 D. Ivanova, J. Barrett, D. Wiedenhofer et al., "Quantifying the Potential for Climate Change Mitigation of Consumption Options," *Environmental Research Letters* 15, no. 9 (2020).

22 Khalfan et al. (2023).

23 Khalfan et al. (2023).

to 14 000 Euros. The carbon footprint of the consuming class is around 7 tCO2e/capita per year, close to the global average. The consuming class is the most prevalent class in the EU, making up more than half of the population, while it represents the second largest class in North America, including a third of the population, according to Ulvila and Wilén's (2017) classification. Furthermore, in China, around a third of the population belongs to this class, exceeding the number of individuals representing the class in the EU and North America combined. India has a small share belonging to the consuming class, below 5% or around 30 million people.

This class contains a diverse group representing vastly different income deciles in various countries. For example, it includes 6.5 million French in the 3rd income decile and 8 million Germans in the 2nd decile, whereas in Finland, only the first decile has footprints around 7 tCO2e/capita per year. In China, this class is represented by people in the richer segment of the population. In poorer countries, a much smaller share of the population belongs to the consuming class.

3.1.2.4 *The Sufficiency Class*

The sufficiency class includes about one third of the human population, constituting the group of people between the consuming class and struggling class. We suggest it could refer to individuals with carbon footprints of up to 5 tCO2e/capita per year. This shifts attention from North America and the EU to lower-income regions. The vast majority of people in this class lives in developing or rapidly growing economies. According to estimates, around half of the population in China represents the sufficiency class. Rich regions have a lower share of sufficiency class. In the EU, around 15% of the population represents the sufficiency class, whereas the share is around 5% in North America and other rich countries.[24]

Ulvila and Wilén define this class as people who "have their basic needs met by livelihoods that do not cause excessive GHG emissions, and their lifestyles do not consume an excessive amount of energy or non-renewable resources." If considering high life expectancy as a measure of having basic needs met, there are countries with life expectancy above 70 years with less than 1 tCO2e/capita per year across regions and within various income ranges. This group includes countries in Latin America, Asia, Eastern Europe, and North Africa with incomes ranging from USD2 500–12 000 per capita. For example, Paraguay, Nicaragua, the Philippines, and Sri Lanka have footprints

24 Ulvila and Wilén (2017: 128).

around 0.3 tCO2e/capita per year, and Albania, Costa Rica, and Egypt have footprints around 0.5 tCO2e/capita per year.[25]

In rural households in China, four out of five income classes have footprints below 2.5t CO_2/capita, whereas the low, lower-middle, and middle classes have carbon footprints of maximum 1t.[26] The emissions of urban households are larger, but nonetheless, three out of five classes have carbon footprints of maximum 2.5 tons, whereas the high-income class is estimated to have a carbon footprint close to 6 tons.

Regarding the EU, the income threshold of 6000 Euros corresponds roughly to the first income quintile with carbon footprints around 3 tCO2e/capita, according to a study by Ivanova and Wood.[27] Depending on how the sufficiency class is defined, a larger share could be included: according to the study, half of the population in the EU has an average footprint of 5 tCO2e/capita per year, spending a large share of their income on basic consumption including food, housing, and other necessities. Countries with 70–90% of households in this category include, for example, Croatia, Romania, Hungary, Poland, Latvia, Bulgaria, and Spain. Also, countries with the largest share of households belonging to the richest 10% have households belonging to the bottom 50%, including Luxembourg, Great Britain, Ireland, and Finland.

Approximately 5% of EU households already have carbon footprints in line with the 2030 climate target. Countries with the largest share of households with footprints below 2.5 tons include Romania (more than half of the households) and Hungary, Croatia, Latvia, and Portugal (every fourth to every fifth household). This raises questions about whether these households have their basic needs met and can be classified as the sufficiency class, or whether part or all of them are rather part of the struggling class. At the other end, the EU countries with the lowest share (less than 1%) of households in line with the climate targets include Germany, Ireland, Greece, and Luxembourg.

3.1.2.5 *The Struggling Class*

The struggling class refers to a third of the global population that does not have its basic needs met, lacking essentials like safe and sufficient food, water, and shelter. The struggling class includes the poorest individuals, mainly living in sub-Saharan Africa, where 70% of the population belongs to this class.[28] The

---

25 J. K. Steinberger, J. T. Roberts, G. P. Peters, and Giovanni Baiocchi, "Pathways of Human Development and Carbon Emissions Embodied in Trade," *Nature Climate Change* 2, no. 2 (2012).

26 L. Wei, C. Li, J. Wang et al., "Rising Middle and Rich Classes Drove China's Carbon Emissions," *Resources, Conservation and Recycling* 159 (2020).

27 Ivanova and Wood (2020).

28 Ulvila and Wilén (2017).

respective share in China is in the magnitude of 10%, whereas in India and other Asian developing countries it constitutes almost half of the population. In addition, Ulvila and Wilén (2017) assume that 5% of the people in industrialized countries belong to this group, including people who are struggling with unemployment, homelessness, and insufficient healthcare.

The struggling class is largest in developing countries. The 10% with the lowest income have carbon footprints around 0.1 tCO2e/capita per year and include countries like Honduras, Mozambique, Rwanda, and Malawi. Their footprints are 2000 times lower than that of top emitters. More than a billion of the poorest people have a purchasing power parity of less than USD 1.90 per day, with carbon footprints up to 1.9 tCO2e/capita per year.[29]

3.1.3 Identifying Carbon Footprint Hotspots (Step 4)

To reduce the environmental footprint, the next step is to identify the most harmful sources, i.e. the 'hotspots.' Household consumption is commonly divided into four categories, housing, nutrition, transport, and goods and services, all of which have considerable climate impacts.

Here, we consider the first three categories and aim to identify single high-carbon activities with large sufficiency-oriented emission reduction potential. More than 60 consumption options to mitigate CO_2 emissions have been identified in an extensive literature review by Ivanova et al. (2020) while carbon-intensive 'hotspots' have been identified by Akenji et al. (2021: 44–55) (see Table 4.3). Related to transport, both identified a need to reduce car and air travel. Related to food, reducing the consumption of animal products, in particular meat and dairy, have high emission reduction potential. Related to housing, switching from non-renewable to renewable electricity, producing own renewable electricity, and construction and maintenance, were identified as having high emission reduction potential.

For this chapter, we focus on three top hotspots that we consider relevant for a large segment of the population in affluent countries and communities with large carbon footprints: air travel, private car use, and meat consumption.

While recognizing the importance of addressing housing, we consider the hotspot of switching to renewable electricity an efficiency rather than sufficiency measure. As the production of own renewable electricity and the involvement in construction and maintenance are not possible for all individuals, they were left out. In the food sector, despite dairy products having considerable climate impact, meat was chosen due to its larger carbon footprint.

29 Chancel and Piketty (2015); K. Hubacek, G. Baiocchi, K. Feng et al., "Global Carbon Inequality," *Energy, Ecology and Environment* 2, no. 6 (2017).

TABLE 4.3 Consumption options to mitigate CO_2 emissions[a]

Reference	Consumption option	Mitigation potential (tCO2eq/cap/y)
Ivanova et al. 2020	Identifying consumption mitigation options	
	Transport	
	– Living car-free	Range: 3.6–0.6
	– Shifting to a battery electric vehicle	Range: 5.4–(−1.9)
	– Air travel, one less flight 1) long return 2) medium return	Ranges: 1) 4.5–0.72; 2) 1.5–0.2
	Food	
	– Shifting to vegan diet	Range: 2.1–0.4
	– Shifting to vegetarian/Mediterranean diet	Range: 2–0.01
	Housing	
	– Purchasing renewable electricity	Range: 2.5–0.3
	– Producing own renewable electricity	Range: 4.8–0.1
Akenji et al. 2021	Identifying'carbon hotspots' 1) high-income countries; 2) upper-middle-income countries; 3) lower-middle-income country	
	Transport	
	– Car	Ranges: 1) 3.54–1.25; 2) 0.72–0.1; 3) 0.34
	– Air travel	Ranges: 1) 1.44–0.57; 2) 0.35–0.1; 3) 0.04
	Food	
	– Meat	Ranges: 1) 1.39–0.33; 2) 1.23–0.09; 3) 0.03
	– Dairy	Ranges: 1) 0.55–0.18; 2) 0.32–0.01; 3) 0.18
	Housing	
	– Non-renewable grid electricity	Ranges: 1) 1.32–0.36; 2) 0.79–0.07; 3) 0.19
	– Construction and maintenance	Ranges: 1) 0.52–0.23; 2) 0.37–0.19; 3) 0.1

a Notes regarding Ivanova et al. (2020): A negative value represents the potential for backfire. Related to the "shifting to a battery electric vehicle (BEV)," the authors note that due to differences in assumed travel distances, living car-free appears to have lower mitigation potential than shifting to BEVs. Regarding Akenji et al. (2021): High-income countries studied included Canada, Finland, UK, and Japan; upper-middle-income countries included China, South Africa, Turkey, Brazil, and Indonesia; and the lower-middle-income country studied was India

3.1.4 Carbon 'Hotspot Footprints' for Different Income Classes (Step 5)

In the following, the hotspots of flying, private car use, and meat consumption are scrutinized to map the carbon footprints of each income class. The numbers are summarized in Table 4.4, including the data sources of this sub-section.

3.1.4.1 *Air Travel*

Air travel is the consumption category that increases the most with rising incomes, and it is a luxury with high carbon emissions. Essentially, it is the overconsuming 10% who fly—and the ultra-rich 1% has by far the highest carbon footprint from flying. Data on consumption patterns of the ultrarich are scarce[30] and therefore are often based on estimates. For example, Chancel and Piketty (2015) estimate the air travel carbon footprint of an ultrarich person to be 35 tCO2e per year, in a possible breakdown of an ultrarich American's total carbon footprint of 300 tCO2e/capita per year. According to calculations of the authors, flying emissions of 35 tons equals, for example, five round trips between New York and Los Angeles and two round trips from the US to Europe. In this example, air travel constitutes 12% of the total footprint. In a similar estimate of a two-person super-rich household, air travel comprises 33 tCO2e/capita per year, constituting half of the total per capita footprint. An estimate of the richest 1% of EU households is 23 tCO2e/capita per year, representing 40% of the total footprint. The footprint of the richest 1% from air travel can be much larger in the case of frequent flyers or for private jet owners.

The rest of the overconsuming class includes a wide span of incomes, and studies provide examples of air travel constituting between less than 1 and almost 10 tCO2e/capita per year, depending on income. Examples from the US include a two-person upper-income household with a 6-ton carbon footprint per capita from flying. Another estimate of a hypothetical US household of the same size and with the same income is 8 tCO2e/capita per year, equaling, for example, eight return flights from London to New York. In a hypothetical lower-income five-person household in the US, one transatlantic flight is made per year, with a carbon footprint of 1 ton. For the richest 10% in the EU, air travel is estimated to be 3 tCO2e/capita per year.

Emissions from flying drop almost to zero in the consuming class. For example, in the mid 40% of the EU, half of which belong to the global overconsuming class and the rest to the consuming class, the footprint from flying is estimated to be less than 0.1 tCO2e/capita per year.

30 Otto et al. (2019).

Air travel is negligible for the global sufficiency class, including, for example, a considerable share of the population in the EU with total footprints averaging 5 tCO2e/capita per year.

3.1.4.2 *Private Car*

As outlined in Table 4.4, estimates of emissions from car use among the richest 1% range from 5 to 12 tCO2e/capita per year. An estimate for this class in the US is 10 tCO2e/capita per year for car emissions. A more conservative estimate is 5 tCO2e/capita per year for a super-rich household of two people. For the richest 1% EU households, an estimated footprint related to private car use is 12 tCO2e/capita per year, constituting a fifth of the total footprint.

For the richest 10% in the US, footprints for car use are estimated to be half of the total footprint of an average person belonging to the richest 1%, around 5 tons. An estimate for the EU is 7 tCO2e/capita per year, constituting a third of the total footprint.

The consuming class in rich countries have footprints of 2 to 3 tCO2e/capita related to car use. For example, a study on EU countries estimates that the mid 40% income class (half of which belong to the overconsuming class and half to the sufficiency class), is estimated to have a car-related footprint of 3 tCO2e/capita per year. In China, the carbon footprint from car and motorcycle use averages slightly below 1 tCO2e/capita per year. This is assumed to apply to the sufficiency class, the value being somewhat higher for the consuming class. The corresponding average value for India is 1.5 tCO2e/capita per year, assumed to apply to the sufficiency class.

Related to the sufficiency class, the lowest income decile in the US is estimated to have car-related emissions of 2 tCO2e/capita per year, including indirect emissions. Since the 1980s, the trend in the US has been the higher use of private transport in poor households. In France, the footprint of car use for the lowest income decile is calculated to be 0.3 tCO2e/capita per year, including direct emissions. For the purpose of this study, we assume that half of the lowest income deciles in the US and France belong to the sufficiency class and half to the struggling class.

3.1.4.3 *Meat*

There are vast differences in meat consumption between nations and income groups.[31] Put in relation to transport, the difference between income classes

31 Y. Li et al. "Reducing climate change impacts from the global food system through diet shifts," *Nature Climate Change* 14, no 9 (2024).

is nonetheless smaller. This can be explained by there being an upper level to how much more a person can eat. As part of a hypothetical footprint of ultrarich 1% Americans, a study allocates 10 tCO_2e/capita per year for eating beef, corresponding to spending 30 dollars a day on this. Related to the rest of the overconsuming 10%, studies indicate carbon or GHG footprints of 2.5–3 tons for the US and Canada and 1 ton for the EU. In another study on Americans, meat consumption was found to be quite identical in households representing the highest and lowest income quantiles, with footprints around 2 tons in a two-person household, translated to around 1 tCO_2e/capita per year. Rich households spend more money on meat because they buy more expensive products. The consuming and sufficiency classes have per capita footprints of meat consumption between less than 0.5 and 1.5 tCO_2e/capita per year, with North America representing the upper end and China and India the lower end.

3.1.5 Applying a Sufficiency Perspective (Step 6)

This step includes assessing the 'hotspot footprints' of each income class in relation to per capita boundaries connected to the 2030 climate targets. The sufficiency perspective represents staying within ecological boundaries while ensuring that basic needs are met. Hence, the assessment reveals whether the footprints of the hotspots are above, at, or below a sustainable level.

Applying a sufficiency perspective points to the disproportionately large footprints of the richest 1% and 10% of the global population in relation to a per capita boundary. Other classes are closer to living within the boundaries. The classes below the social foundation need support to meet basic needs.

If referring to sufficiency as staying within ecological and social boundaries, the picture becomes rather multi-faceted for rich countries. Related to ecological boundaries, only 5% of EU households have footprints of maximum 2.5 tons, in line with the 2030 climate targets.[32] The social threshold, again, can be addressed through national 'reference budgets', which outline a minimum level of consumption for being able to participate in a certain society at a certain time. In Finland, for example, the reference budget translates into a carbon footprint of 5 tons, half the size of an average Finn, which is still twice as large compared with the 2030 climate target.[33] Other studies point in the same direction. Among minimum income receivers in Finland, the material footprint was almost half that of an average Finn, but only a homeless

32 D. Ivanova and R. Wood (2020).

33 T. Nyfors et al., "Ecological Sufficiency in Climate Policy," *Futura*, no. 3 (2020).

person had a sustainable level of resource use.[34] A UK study regarding the social minimum consumption level found that the household type with the lowest emissions had a footprint that was less than two thirds of an average British household, yet still exceeding the 2030 per capita boundary by a factor of three.[35]

An intriguing question related to sufficiency proposals concerns the size of the footprint of the sufficiency class, perhaps particularly interesting for affluent countries. And, connected to this, how large a share of the emission reductions can be attributed to individual choices and to structural changes, respectively?

That even the carbon footprint of minimum consumption exceeds the ecological boundary in rich countries illustrates the magnitude of the challenge. Similarly, for an EU or North American citizen, it may appear challenging to picture an annual income level below 6000 Euros, which is the income for the sufficiency class in these regions. Still the fact remains, average carbon footprints in North America and many EU countries are manifold compared with the 2030 per capita target level. Given the strong correlation between income and carbon footprint, it seems necessary to address income levels.

To open rather than close doors to sufficiency, i.e. making it feasible rather than making it seem impossible, we suggest that the reference budget, or something like it, could define the upper footprint boundary of the sufficiency class. For the purpose of this study, we use a level around 5 tons, which is supported by the Finnish reference budget along with the average footprint of the EU lower-income half of the population, with emissions largely connected to necessities. For people already living on a minimum it could be reasonable to assume that emission reductions would be attained mainly through structural changes and changing societal norms. This contrasts with the richest individuals, where focus is strongly on behavioral change.

To visualize what living in accordance with the 2030 climate targets could mean in practice, the consumption of people with an average footprint of 5 tCO2e/capita per year, in the study by Ivanova and Wood (2020), can serve as an example. The lion's share of the income is spent on basic consumption—food stands for 1 ton, housing 1.8 tons, whereas the rest, approximately 2.2 tons, is comprised of the consumption areas of travelling by land, clothing, services, and manufactured products. When linking this to the carbon hotspots, it is evident that there is not much space for flying and a limited possibility for car use

34 T. Hirvilammi et al., "Studying Well-Being and Its Environmental Impacts," *Journal of Human Development and Capabilities* 14, no. 1 (2013).

35 A. Druckman, and T. Jackson, "The Bare Necessities," *Ecological Economics* 69, no. 9 (2010).

and eating meat. To ensure that needs are met for all within ecological boundaries, we assume that the per capita footprint from flying needs to be zero or close to zero. With regard to land travel, assuming that it constitutes half of the 2.2 tons above, the carbon footprint is around 1 ton. And with regards to meat, we make the rough assumption that it needs to be less than 1 ton because that is the amount of emissions from food as a whole.

As can be seen above, a needs-versus-wants perspective appears rather naturally in the sufficiency perspective—needs are given priority over wants. In the case of the three carbon hotspots, neither air travel, private car use, nor meat, are by definition needs, but what Gough (2017) refers to as 'need satisfiers.' For example, food is clearly a need, whereas the same cannot be said categorically about meat; a private car may be necessary or not depending on societal structures, and flying, particularly long-haul holiday travel, is predominantly a want. Using a per capita boundary of up to 5 tons per person still provides some space for various choices.

3.1.6 Establishing Sufficiency Proposals for Different Income Classes, Focusing on Carbon Hotspots (Step 7)

To establish sufficiency proposals for the different income classes in accordance with the 2030 climate targets, we focus on hotspots and use reductions of 90%, 75%, and 50% to express the magnitude of change required. In addition, a 'minor reduction' refers to a less than 50% reduction, for example a 10–20% reduction. For an overview of sufficiency proposals for the income classes, see Table 4.4.

The proposal for the richest 1% with an extremely high carbon footprint is characterized by immediate and drastic reductions regarding all hotspots. It implies at least a 90% reduction in air travel, combined with policy for phasing out private jets and downscaling long-haul tourism. Sufficiency proposals for the richest also include a 75–90% reduction in private car use and meat consumption.

Immediate reductions are equally required by the rest of the richest 10% with very high carbon footprints. The reduction requirements are enormous, but the numbers are not as astronomical as for the richest 1%. A reduction of 75–90% is needed in air travel and private car use, with current CO_2 emissions of the same magnitude. Also, meat consumption needs to be halved.

For the consuming class, more modest reductions are proposed. The most significant hotspot is private car use, where a 50% reduction is proposed. Similarly, minor reductions are proposed in meat consumption. People in the consuming class fly very little or not at all, hence the proposal includes not increasing flying or not starting to fly, but to keep traveling by land. Ulvila and

TABLE 4.4 Sufficiency-based proposals according to income classes (the data sources are indicated in square brackets and listed in the footnote)[a]

Income class / Hotspot	Ultrarich over-consuming (richest 1%)	Over-consuming (richest 10%)	Consuming (18%)		Sufficiency (40%)	Struggling (32%)	
			Rich countries (6%)	China, India, and other regions of the world (12%)		Rich countries (1%)	China, India, and other regions of the world (31%)
Air travel							
Footprint (tCO2e/person per year)	US: 33 [6], 35 [4] EU: 23 [7]	US: 6 [1], 0.5–8 [5] EU: [illegible] [7]	EU: <0.1 [7]		Negligible [7]	–	–
Proposal	Minimum 90% reduction	75–90% reduction	Minor reduction. Do not increase or start flying, keep traveling by land		Do not start flying, keep traveling by land.	Policy: Improve affordable access to public transport	Policy: Enable access to needs-based travel
Private car							
Footprint (tCO2e/person per year)	5–12 US: 5 [6], 10 [4] EU: 12 [7]	5–7 US: 5 [2] EU: 7 [7]	2–3 [4,7]	1–2 China: 1 [8] India: 2 [8]	<1–2 US: 2 [2] EU: <1 [7] China: 0.7 [8] India: 1.5 [8]	–	–
Proposal	75–90% reduction	75–90% reduction	50% reduction	50% reduction	Keep living car-free or driving very moderately	Policy: Improve affordable access to public transport	Policy: Enable access to public transport

Meat							
Footprint (tCO_2e/person per year)	10 [4]	1–3 US: 1 [1], 2.5 [9] Canada: 3 [9] EU: 1 [9]	0.5–1.5 US & Canada: 1.5 [9] EU: 0.5–1 [9]	<0.5–1 China: 0.5–1 [9] India: <0.5 [9]	0–1 US: 0.5 [9], 1 [1] EU: 0.5 [9] China: 0.5 [9] India: 0–0.5 [9]	0–0.5 [9]	0–0.5 [9]
Proposal	75–90% reduction for high-meat diet	50% reduction	Minor to 50% reduction	Minor reduction	No or minor reduction	Policy: Enable access to predominantly vegetarian or vegan food which fulfills nutritional requirements	Policy: Enable access to predominantly vegetarian or vegan food which fulfills nutritional requirements

a The average carbon footprints of the hotspots are outlined for different income classes (tCO_2e/capita per year). This chapter proposes reductions in the magnitude of 90%, 75%, and 50% by 2030. 'Minor reduction' refers to less than 50% reduction.

Data sources: [1] Jones and Kammen (2011); [2] Chancel (2014); [3] Ummel (2014); [4] Chancel and Piketty (2015); [5] Lacroix (2018); [6] Otto et al. (2019); [7] Ivanova and Wood (2020); [8] Akenji et al. (2021); and [9] Li et al. (2024)

Wilén (2017) suggest that measures for this class could be taken after the richest have been addressed.

The sufficiency class, more than one third of the global population, can continue as before or with minor lifestyle changes. The point is to not start flying, but to keep traveling by land, and to continuing living car-free or driving very moderately. Regarding meat, a minor (10–20%) reduction in meat consumption may be recommended depending on other lifestyle choices.

The proposal regarding the struggling class in poor countries, one third of the global population, is empowerment. We also include policy intervention to enable access to nutritionally good, predominantly vegetarian or vegan food and improved access to public transport. The share of the struggling class is small in rich countries, and poverty is rarely a question of scarcity but primarily a question of distribution. The per capita footprint of the struggling class in rich countries may be larger than the population average of lower-income countries. This points to the particular importance of a sufficiency transformation at the societal level. This includes, for example, supporting shifts in societal values, to prevent a low-consumption lifestyle, where basic needs are met, from being socially insufficient and constituting a limiting factor for participating in society. With a societal shift toward sufficiency and structural changes to support a low-consumption lifestyle, segments of the struggling class in rich countries could perhaps classify as part of the sufficiency class.

It may appear contradictory from an equality and justice perspective to suggest emission reduction magnitudes that, to some extent, allows the rich to continue their harmful consumption patterns while restricting access to these consumption options for the poor. These suggestions for 2030 are outlined as an intermediate target; we suggest considering per capita footprint targets in line with limiting global warming to 1.5 degrees for each decade until 2050: 1.4 tCO2e/person per year by 2040 and 0.7 tons by 2050, as outlined by Akenji et al. (2021). Hereby, the sufficiency level is a changing target, and sufficiency-based proposals could include stepwise further reductions in line with these. The question to what extent the poor have the right to gain access to 'hotspot' consumption options is beyond the scope of this chapter. As referred to in this chapter, sufficiency is about staying within ecological boundaries while ensuring that needs are met. Simultaneously, the notions of needs and wants ought to be discussed.

Societies and individuals can respond to this by structural and behavioral changes. Even though the richest can and should contribute, voluntary action is likely not enough—policy intervention is necessary in the societal sufficiency transformation.

4 Implications

4.1 *Sufficiency-Based Proposals Based on Multiple Footprint Hotspots*

The seven analytical steps for establishing sufficiency-based proposals, above applied to the carbon footprint, can equally be implemented to other footprints. Then previously hidden areas of consumption may become visible, or the magnitude of importance can shift between hotspots. An optical illusion card can perhaps serve as an analogy—depending on from what angle it is viewed, various dimensions stand out, i.e. different footprint hotspots are more dominant. Below, we briefly consider the water and material footprints.

The water footprint relates to the planetary boundary of freshwater use, indispensable for ecosystems and human societies. The water footprint is predominantly caused by agricultural production. This directs a particular focus to food. The major contributors to the water footprint are cereal, meat, and milk.[36] Meat is particularly interesting from a sustainability as well as a sufficiency and needs-versus-wants perspective, as it can be replaced by other protein alternatives. Hence, meat is a major hotspot when the water footprint is the chosen lens. Proposals for action would place a strong emphasis on reducing meat consumption in high-consuming classes, but also on the other classes particularly in rich countries.

The material footprint covers a broad range of resources. According to the International Resource Panel IRP (2019), main raw materials include biomass, fossil fuels, metal ores, and non-metallic minerals. The material footprint does not have a single corresponding planetary boundary but links closely to, for example, biosphere integrity, climate change, water scarcity, biogeochemical flows, and earth surface change.[37] The material footprint is high for the consumption areas of housing, mobility, and nutrition. Housing and construction stand out as particularly material intensive. It can, for example, constitute more than 40% of the total material footprint of an average person living in the Brussels capital region.[38] When considering the material footprint, results point to the importance of addressing housing. On an individual level, proposals would, for example, imply reducing the number of square meters of living space and considering building materials.

---

36 M. Mekonnen and A. Y. Hoekstra, *National Water Footprint Accounts* (UNESCO, 2011).

37 Vanham et al. (2019).

38 M. Christis, A. Athanassiadis, and A. Vercalsteren, "Implementation at a City Level of Circular Economy Strategies and Climate Change Mitigation—the Case of Brussels," *Journal of Cleaner Production* 218 (2019).

With these brief examples, we aim to highlight that the hotspots and corresponding actions look different depending on which footprint is considered. For example, solely considering the carbon footprint underestimates the role of food in efforts to live within the planetary boundaries; the importance of addressing diets becomes emphasized when considering the water footprint. When considering water footprint hotspots, again, the large environmental impact of air travel is invisible. Hence, 'reduce flying' would not emerge as an important category when considering only the water footprint. Replacing cars running on fossil fuels with electric cars can serve as another example. In a carbon footprint perspective, an electric car has lower emissions, but when considering the material footprint, there is no considerable difference.[39] The material footprint further points to the complexity in how human activities relate to multiple harmful environmental impacts. Taken together, it is of crucial importance to consider multiple footprints when designing sufficiency-based actions and policies.

4.2 *New Role Models*

Perhaps one of the most crucial take-aways of formulating class-based sufficiency proposals is that a considerable share of the global human population belongs to the sufficiency class. These persons can be in a position of lifestyle role models, experts to learn from, and contemporary examples to be inspired by. In the words of Trebeck and Williams (2019), they have 'arrived.' Previously, the sufficiency class has largely been ignored, as the focus has been on the richest and poorest; however, as Ulvila and Wilén (2017) put it, it is high time for this part of the population to receive the positive attention they deserve.

Bringing forth role models might be a useful way to promote sufficiency lifestyles. This would imply highlighting, for example, how some societies are organized with a high share of public transport, and how nutritional, vegetarian diets can be learned from elsewhere. This means, for example, that India can serve as a role model regarding food. In the EU, there is perhaps reason to look more closely at its own sufficiency class. Are there lessons on how society works in, for example Croatia, Poland, and Spain, regarding the moderate per capita carbon footprints, which can be learned by other countries? Also, questions arise regarding to what extent needs are met for the sufficiency class, and whether there are factors preventing full participation in society. Without downplaying crucial questions related to social sustainability, perhaps changing societal norms in a sufficiency direction ought to be a focus rather than rising consumption levels? Politically these are, however, challenging topics,

39 M. Lettenmeier, C. Liedtke, and H. Rohn, "Eight Tons of Material Footprint," *Resources* 3, no. 3 (2014).

and questions arise related to how they could compete electorally in a context of standard political rhetoric centered around growth and jobs.

For a societal shift to sufficiency, a deep cultural transformation is needed. This includes addressing vast questions such as a transformation of the growth-based economic system, the logic of business, and preventing a 'sufficiency rebound.' Connected to this are questions of power, vested interests, corruption, and short-sighted politics. The relevance of these questions, even though not addressed in greater detail here, cannot be over-emphasized. Unless society at large supports a sufficiency transformation, role models make a limited difference. More broadly, further research is also needed regarding how the different footprints are linked and overlap in terms of planetary boundaries.

4.3 *Towards 'Enough' with Sufficiency-Based Proposals*

To address the planetary emergencies, elements of sufficiency are increasingly called for to complement prevailing efficiency strategies, which have failed to reduce overall human environmental pressure. While recognizing the importance of addressing all three factors of the IPAT equation, from our perspective of living in an affluent country context, we want to emphasize the responsibility of looking at ourselves in the mirror and argue that affluence, and connected to it, technology, necessitates the most attention.

This chapter develops sufficiency-based proposals for income classes by linking the footprint, planetary boundary, and sufficiency concepts. The carbon footprint is chosen to inform of the sufficiency approach. Identified carbon footprint hotspots include air travel, private car use, and meat consumption. The footprints of the richest attract particular attention. Proposals for the richest 1% imply an immediate and drastic reduction in air travel, private car use, and also in meat consumption. Air travel requires the largest reduction—per capita emissions from flying alone can equal more than ten times the 2030 per capita climate target. For the richest 10%, immediate, albeit not fully as drastic, reductions are required in the same consumption areas.

Utilizing an income class perspective to operationalize sufficiency highlights the need for different sufficiency strategies for different classes. Also, it points to the fact that a sufficiency class already exists. In times of planetary emergencies, people of the sufficiency class are the (new) role models. The seven analytical steps for operationalization, here illustrated through the carbon footprint, can be applied to other footprints. This may point to other hotspots, which contributes to formulating comprehensive sufficiency-based proposals.

A broad discussion on what is 'enough,' and on needs versus wants is urgently needed. We suggest that sufficiency-based proposals for income classes, considering multiple footprints, could support the transformation of shifting from current levels of global overconsumption to living within the planetary boundaries.

Acknowledgements

We would like to thank Pasi Heikkurinen for discussions held during the process and reviewers for useful comments on earlier drafts of this chapter. The work of Tina Nyfors was supported by the Tiina and Antti Herlin Foundation.

References

Akenji, L. et al. (2021). "1.5-Degree Lifestyles: Towards A Fair Consumption Space for All." Berlin: Hot or Cool Institute.

Alcott, B. (2008). "The Sufficiency Strategy: Would Rich-World Frugality Lower Environmental Impact?" *Ecological Economics* 64(4): 770–786.

Biermann, F., and Kim, R.E. (2020). "The Boundaries of the Planetary Boundary Framework: A Critical Appraisal of Approaches to Define a 'Safe Operating Space' for Humanity", *Annual Review of Environment and Resources*, 45(1): 497–521.

Chancel, L. (2014). "Are Younger Generations Higher Carbon Emitters than Their Elders?" *Ecological Economics* 100 (April): 195–207.

Chancel, L., and Piketty, T. (2015). "Carbon and Inequality: From Kyoto to Paris." Paris School of Economics.

Christis, M., Athanassiadis, A., and Vercalsteren, A. (2019). "Implementation at a City Level of Circular Economy Strategies and Climate Change Mitigation—the Case of Brussels." *Journal of Cleaner Production* 218: 511–520.

Credit Suisse Research Institute. (2022). "Global Wealth Databook 2022." Credit Suisse Research Institute.

Downing, A.S., *et al.* (2020). "Learning from generations of sustainability concepts," *Environmental Research Letters* 15: 1–14.

Druckman, A., and Jackson, T. (2010). "The Bare Necessities: How Much Household Carbon Do We Really Need?" *Ecological Economics* 69(9): 1794–1804.

Elhacham, E. et al. (2020). "Global Human-Made Mass Exceeds All Living Biomass." *Nature* 588(7838): 442–444.

Gough, I. (2017). *Heat, Greed and Human Need: Climate Change, Capitalism and Sustainable Wellbeing*. Edward Elgar Publishing.

Hayden, A. (2020). "Sufficiency." In *Routledge Handbook of Global Sustainability Governance*, edited by A. Kalfagianni, D. Fuchs, and Anders Hayden. Routledge.

Heikkurinen, P. (ed.) (2017). *Sustainability and Peaceful Coexistence for the Anthropocene*. Routledge.

Hirvilammi, T. et al. (2013). "Studying Well-Being and Its Environmental Impacts: A Case Study of Minimum Income Receivers in Finland." *Journal of Human Development and Capabilities* 14(1): 134–154.

Hubacek, K. et al. (2017). "Global Carbon Inequality." *Energy, Ecology and Environment* 2(6): 361–369.

IPBES (2019). *Summary for Policymakers of the Global Assessment Report on Biodiversity and Ecosystem Services of the Intergovernmental Science-Policy Platform on Biodiversity and Ecosystem Services*. IPBES secretariat.

IRP. (2019). *Global Resources Outlook 2019: Natural Resources for the Future We Want*. United Nations Environment Programme.

Ivanova, D. et al. (2020). "Quantifying the Potential for Climate Change Mitigation of Consumption Options." *Environmental Research Letters* 15(9): 093001.

Ivanova, D., and Wood, R. (2020). "The Unequal Distribution of Household Carbon Footprints in Europe and Its Link to Sustainability." *Global Sustainability* 3:e18.

Jones, C. M., and Kammen, D.M. (2011). "Quantifying Carbon Footprint Reduction Opportunities for U.S. Households and Communities." *Environmental Science & Technology* 45(9): 4088–4095.

Jungell-Michelsson, J., and Heikkurinen, P. (2022). "Sufficiency: A Systematic Literature Review." *Ecological Economics* 195 (May):107380.

Khalfan, A. et al. (2023). *Climate Equality: A Planet for the 99%*. Oxfam International.

Lacroix, K. (2018). "Comparing the Relative Mitigation Potential of Individual Pro-Environmental Behaviors." *Journal of Cleaner Production* 195 (September): 1398–1407.

Lettenmeier, M., Liedtke, C., and Rohn, H. (2014). "Eight Tons of Material Footprint—Suggestion for a Resource Cap for Household Consumption in Finland." *Resources* 3(3): 488–515.

Li, Y., He, P., Shan, Y., Li, Y., Hang, Y., Shao, S., Ruzzenenti, F., and Hubacek, K. (2024). "Reducing Climate Change Impacts from the Global Food System through Diet Shifts." *Nature Climate Change* 14(9): 943–953.

Matuštík, J., and Kočí, V. (2021). "What Is a Footprint? A Conceptual Analysis of Environmental Footprint Indicators." *Journal of Cleaner Production* 285 (February): 124833.

Mekonnen, M. M., and Hoekstra, A.Y. (2011). *National Water Footprint Accounts: The Green, Blue and Grey Water Footprint of Production and Consumption*. 50. Value of Water Research Report. Unesco-IHE Institute for Water Education.

Nyfors, T. et al. (2020). "Ecological Sufficiency in Climate Policy: Towards Policies for Recomposing Consumption." *Futura*, no. 3.

O'Neill et al. (2018). "A Good Life for All within Planetary Boundaries." *Nature Sustainability* 1(2): 88–95.

Otto, I. M. et al. (2019). "Shift the Focus from the Super-Poor to the Super-Rich." *Nature Climate Change* 9(2): 82–87.

Oxfam. (2015). *Extreme Carbon Inequality*. Oxfam.

Pathak, M. et al. (2022). "Technical Summary," in *Climate Change 2022: Mitigation of Climate Change. Contribution of Working Group III to the Sixth Assessment Report of the Intergovernmental Panel on Climate Change*, Shukla, P. et al. (eds). Cambridge University Press.

Planetary Boundaries Science (PBScience) (2025) *Planetary Health Check 2025*. Potsdam, Germany: Potsdam Institute for Climate Impact Research (PIK).

Princen, T. (2005) *The Logic of Sufficiency*. Cambridge: MIT Press.

Richardson, K. et al. (2023). "Earth beyond Six of Nine Planetary Boundaries." *Science Advances* 9(37): eadh2458.

Rockström, J. et al. (2009a). "A Safe Operating Space for Humanity." *Nature* 461(7263): 472–475.

Rockström, J. et al. (2009b). "Planetary Boundaries: Exploring the Safe Operating Space for Humanity." *Ecology and Society* 14(2).

Spengler, L. (2016). "Two Types of 'Enough': Sufficiency as Minimum and Maximum." *Environmental Politics* 25(5): 921–940.

Steinberger, J. K. et al. (2012). "Pathways of Human Development and Carbon Emissions Embodied in Trade." *Nature Climate Change* 2(2): 1–5.

Trebeck, K., and Williams, J. (2019). *The Economics of Arrival: Ideas for a Grown-up Economy*. Policy Press.

Ulvila, M., and Wilén, K. (2017). "Engaging with the Plutocene: Moving towards Degrowth and Post-Capitalist Futures." In *Sustainability and Peaceful Coexistence for the Anthropocene*, edited by Pasi Heikkurinen. Routledge.

Ummel, K. (2014). "Who Pollutes? A Household-Level Database of America's Greenhouse Gas Footprint." *SSRN Electronic Journal*.

UNEP. (2020). *Emissions Gap Report 2020*. Nairobi.

United Nations. (2015). "Paris Agreement."

United Nations. (2022). "Convention on Biological Diversity."

Vanham, D. et al. (2019). "Environmental Footprint Family to Address Local to Planetary Sustainability and Deliver on the SDGs." *Science of The Total Environment* 693 (November): 133642.

Wei, L. et al. (2020). "Rising Middle and Rich Classes Drove China's Carbon Emissions." *Resources, Conservation and Recycling* 159 (August): 104839.

Wiedmann, T., and Minx, J. (2008). "A Definition of 'Carbon Footprint.'" In *Ecological Economics Research Trends*, edited by C. C. Pertsova, 1–11. Hauppauge NY: Nova Science Publishers.

World Bank. (2023). "Population, Total." Accessed March 22, 2023. https://data.worldbank.org/indicator/SP.POP.TOTL.

World Meteorological Organization (2025) *State of the Global Climate 2024*. 1368.

Wu, L. et al. (2021). "A Planetary Boundary-Based Environmental Footprint Family: From Impacts to Boundaries." *Science of The Total Environment* 785 (September): 147383.

CHAPTER 5

Affluence at the Expense of Others: from the Hegemony of Capitalism to Degrowth Transitions towards Sufficiency

Toni Ruuska

> I have long thought that the most pessimistic view is one that hopes for the survival of modernity in something like its present form.
>
> JASON W. MOORE, *Capitalism in the Web of Life* (2015, 87)

1 Introduction

My son has a "friend book," where friends and relatives are asked to answer a set of questions ranging from a simple "*Where were you born?*" to the more personal and abstract question "*What do you dream of?*" It is a tender and innocuous way to get to know people, but also to ask questions that you would not perhaps normally ask. When filling out the book on her visit, my now 89-year-old grandmother (my son's great-grandmother) pondered hard on one question that said, "Did you know that I wonder about?" After some time, she said "I wonder how much stuff there is in the stores and where does it all come from?"

This is not the first time my grandmother and I have been united by a shared sense of wonder. Anyhow, there is no straight-forward answer to her question. The world economy is so vast and complex, the supply chains and markets are trans-continental—it is a truly global operation! Up until the Covid-19 pandemic, the lived experience of my grandmother, which roughly coincides with the period of the Great Acceleration,[1] had been that the stores had started to become bigger in size, more plentiful in numbers, and containing ever more stuff. Privileged as I am, I too (though less than half of her age while writing

1 W. Steffen, W. Broadgate, L. Deutsch et al., "The Great Acceleration," *Anthropocene Review* 2, no. 1 (2015).

 | DOI:10.1163/9789004746404_006

this) have experienced this overwhelming abundance here in the overly affluent North. Yet, it is another thing entirely to unveil the curtain from what remains hidden for most. Definitely more stuff, but where does it all come from?

A humble point is to note first that the stuff comes from *somewhere*. Everything that is produced and consumed has a material footprint and environmental consequences.[2] In all cases, the stuff comes from the Earth—from the fields, meadows, soil, mountains, deserts, mines, grasslands, forests, rivers, lakes, seas, oceans, and glaciers—and is produced by the labor of countless workers around the world. The places from which our stuff comes from are habitats. Someone and something (if not a human being, then a non-human being) lives where a certain resource is taken from. The contemporary world economy, driven by economic growth, competition, and the accumulation of capital, relies on the exploitation, extraction, and commodification of these habitats.[3] Growing affluence, or the rising material standard of living, requires the molding, manipulation, and most often, the destruction of these places, coupled with requiring an exploited workforce and/or confined and domesticated species as parts and particles of commodity production and the accumulation of capital.[4]

This chapter discusses how the Global North and industrial centers predominately take advantage of subordinate peoples, landscapes, and habitats to secure and maintain their privileged position in the world economy.[5] It is asserted, based on Alf Hornborg's work, that the capitalist world economy represents a zero-sum game[6] where the "winners" generate their affluence at the expense of others. In addition to unprecedented socio-economic inequality, the ongoing and escalating ecological crisis testifies to the annihilation of non-human beings as, for instance, the processes of extractivism bulldoze rainforests, rattle the ocean floors, dismantle mountains, and cancel out the remaining places of wilderness.[7] Instead of seeking affluence at the expense of

2 See, e.g., E. Pineault, *A Social Ecology of Capital* (Pluto Press, 2023).

3 See, e.g., J. B. Foster, *Marx's Ecology* (Monthly Review Press, 2000; J. W. Moore, *Capitalism in the Web of Life* (Verso, 2015); K. Saito, *Marx in the Anthropocene*, (Cambridge University Press, 2023).

4 M. Arboleda, *Planetary Mine* (Verso, 2020); Pineault (2023); Saito (2023).

5 A. Hornborg, "Zero-Sum World," *Int. J. of Comparative Sociology* 50, n. 3–4 (2009); C. Dorninger, A. Hornborg, D. J. Abson et al., "Global patterns of ecologically unequal exchange," *Ecological Economics*, 179 (2021).

6 A. Hornborg, *Global Magic* (Palgrave Macmillan, 2016); Hornborg (2009).

7 A. A. Dunlap and J. Jakobsen, *The Violent Technologies of Extraction* (Springer, 2020); Arboleda (2020).

others, there is a call for the redistribution of wealth and resources, and a call for the *degrowth* of the overly affluent and overly technologized societies in order for them to move towards sufficiency and sustainability.

The chapter is structured in the following way: I begin with discussing the hegemony of capital, and then I explore ecologically unequal exchange to highlight how the capitalist world economy benefits some at the expense of the rest. I then briefly outline general ideas of degrowth before touching on the concept of sufficiency. In the final section, affluence and technology are discussed from the perspective of degrowth debates to find pathways towards local and communal affluence and less resource-intensive technology.

2 The Hegemony of Capital

Capital does not do anything on its own. It needs agents and institutions to put it into motion and to secure further accumulation. Some people, corporations, and states benefit greatly from the workings of the global capitalist economy while the rest of the humans and non-humans are exploited, dispossessed, or killed. The economy is rigged in such a way that the producers always have an incentive to produce more and further exploit the workers. At the same time, people are persuaded and encouraged to consume more through advertising and marketing, and the politicians are re-elected if the economy keeps on growing. Most national economies are built around the notion of perpetual economic growth and, in the aggregate, so is the contemporary world economy. Thus, there must be more of everything (credit, loans, investments, production, consumption) for the system to reproduce.

Consequently, a focal condition for capital's expansion and reproduction is the perpetual offering of investment opportunities for a variety of economic actors to tap into: capital needs to circulate in order for it to yield returns, for instance, in the shape of new products and services. As I have written elsewhere,[8] technological development is, in our time, perhaps the target of most capital investments, an outlet through which the accumulation of capital is secured. In complex, capital-intensive, and technologized modern societies, capital and technology are symbiotic things; they are conditioned and defined by each other. To develop technology, capital is needed, and to accumulate capital, investment opportunities, new products, and new services are needed.

However, capital's relation to technology is of a certain kind. While in the past technology has been developed for many reasons and purposes,[9] capital

8 T. Ruuska, "Conditions for Alienation," In *Sustainability beyond Technology* (2021).

9 See, e.g., T. Ingold, *Perception of the Environment* (Routledge, 2000); Hornborg (2016).

shapes technology and steers technological development to accumulate more capital. For capital, technology is an investment opportunity and a source of profit, a commodity, a source of value. When up and running, the alliance of capital and technology is an accelerating and expanding juggernaut that is always searching for new markets, without any other purpose than value creation and market dominance.

We know very well that the human-induced impacts on the environment have surged, especially since the 1950s.[10] Now, it would be too straightforward to claim that the reason is solely the unholy alliance of capital and technology, which contributes greatly to the growing world economy, yet it is arguably one of the main reasons (a root cause) for the surge. The alliance also explains why the situation is only getting worse, despite serious interventions (think of, for instance, *The Limits to Growth* report in 1972, the foundation of various environmental NGOs [e.g., Greenpeace, WWF], COP meetings [since 1995] and IPCC reports [e.g., *The Sixth Assessment Report* in 2021], environmental conflicts, climate marches and school strikes, the rise of Extinction Rebellion, and so forth).

To be sure, this is not because there exists a paradox in the organization of capital. Capitalism works according to its logic, which mounts to stratospheric affluence for some humans but to immiseration for most and increasingly catastrophic impacts on the biosphere. It would be schizophrenic to expect that an organization based on expansive capital accumulation and perpetual economic growth would deliver sustainability and preserve habitats for future generations. It has not done it, and it will not do it. Similarly, if capital accumulation steers the development of technology, then the agents of capital will most likely turn technology into a "fetish of doom," as Walter Benjamin claimed,[11] because the principal drive is for expansive capital accumulation, entailing accelerating extraction and exploitation of habitats.

3 A Zero-Sum World

In the zero-sum world,[12] the affluence of the few is secured by the exploitation of unpaid labor and cheap access to resources and technology, especially in the Global South. It is true that capitalism can be regulated. From a socioeconomic standpoint, in the Nordic countries we have seen that the benefits of economic growth were more equally divided during the twentieth century.

10 Steffen et al. (2015).
11 Quoted in E. Traverso, *Revolution* (Verso, 2021), p. 52.
12 Hornborg (2009; 2016).

But although there is evidence that capitalism can be regulated regionally, this does not mean that capitalism is, or can be, just or sustainable globally. The some-over-many composition of capitalism is due to its class-based organization to create value: some own the means of production and make use of others' labor power to produce surplus value. It is an organization of disparity *par excellence*.

Furthermore, productivist and growth-based systems, like capitalism, make use of "nature's free gifts" for the expansive accumulation of capital through the exploitation of human labor power.[13] Some individuals, classes, organizations, and states have benefitted greatly from the exploitation and destruction of natural habitats, and others have been exploited and have suffered deeply. This is our zero-sum world.[14] The scholarly work behind this statement is linked to the ecologically unequal exchange theory and framework that states that, in the predominant capitalist world economy, wealthier, more powerful countries and corporations in the Global North exploit the places that are mainly located in the Global South.[15] This is because the wealthier and more powerful actors have disproportionate access—via regimes of the political, economic, military, legal, and media levers of power—to natural resources and sink capacity. The resources and sinks are interlinked to all stages of global commodity chains—including extraction, production, distribution, consumption, and disposal—which are again linked in the unequal distribution of benefits and negative environmental impacts.[16]

During the past decades international trade volumes have surged as the local demand for materials, energy, land, and labor have increasingly been met by the utilization of non-domestic variants. The theory of ecologically unequal exchange explores the dynamics and quantities of world trade as it explicitly considers the material aspects of international trade and argues that it is based on the asymmetric net transfers of resources from peripheries to the core areas in the global capitalist economy. The high income nations, which form the core of the global capitalist organization, depend on resource-intensive

13 K. Marx, *Capital: A Critique of Political Economy, Volume 1: The Process of Capitalist Production* (Penguin, 1976); Foster (2000); Moore (2015); Saito (2023).

14 Hornborg (2009; 2016).

15 The Global North and Global South should be understood here as analytic categories that inform us about the historical and existing disparities and exploitation in the world economy. While it is true that the North exploits the South, it is also true that there exist great disparities within the North, between its regions and nations.

16 See, e.g., A. Hornborg, "Toward an ecological theory of unequal exchange," *Ecological Economics* 25, (1998); Hornborg (2009); J. G. Givens, X. Huang and A. Jorgenson, "Ecologically unequal exchange," *Sociology Compass* 13, no. 5 (2019).

industrial technologies and infrastructures whose maintenance, functioning, and manufacturing depend on the flow of net transfers of resources from distant parts of the world.

The high-income nations and organizations obtain much higher revenues for the resources they export due to their position in global supply chains and the world economy. As Dorninger et al. (2021, 2) wrote, "the asymmetry of international trade, i.e., of the net transfers of resource volumes and monetary values, is a crucial determinant of the capacity of individual nations to accumulate capital and technological infrastructure and to thereby achieve economic growth." These unequal patterns of trade are the result of global socio-economic inequalities and also reproduce them and stand in the way of socio-ecological sustainability, particularly because of environmental burden-shifting to less affluent nations and regions.

A compelling empirical study by Dorninger et al. published in *Ecological Economics* (2021) includes 170 countries in its analysis and incorporates 99.2% of the world population (in 2015) along with most global supply chains and resource flows. The study explicitly states that, in the contemporary world economy, biophysical resources flow from richer countries to poorer countries in an asymmetric fashion. The authors summarize their findings, which support the foundational premises of the ecologically unequal exchange theory, in the following way:

- High-income nations are net importers of embodied materials, energy, land, and labor.
- High-income nations gain a monetary trade surplus via this resource appropriation.
- Lower-income nations provide resources but experience monetary trade deficits.
- The observed inequality is systemic and hampers global sustainability in multiple ways.

The study portrays a period of 26 years, from 1990 to 2015. By comparing the monetary exchange of resources embodied in trade, significant international disparities were found in how resource provisioning was compensated. It stated that value added per ton of raw material embodied in exports was 11 times higher in high-income countries than it was in those with the lowest income in the global economy and it was 28 times higher per unit of embodied labor. With only China and India as exceptions, every other world region served as net exporters of embodied materials, energy, land, and labor to high-income countries during the investigated period. In general, the processes and structures of ecologically unequal exchange in the capitalist world economy make it possible for high-income countries to simultaneously appropriate resources from distant places and generate a monetary surplus through international trade, which at the same time undermines the position and prospects of the less affluent nations.

If the global capitalist organization is unjust and unpaid for most humans, it is deadly to an increasing portion of non-humans. We already know too well what the state-of-the-art natural sciences are reporting: climate change, biodiversity loss and mass extinction, soil erosion, melting glaciers, ocean acidification, declining fish stocks, plastic islands, eroding coral reefs, deforestation, loss of wilderness, and so forth. Expansive capital accumulation and the economics of perpetual growth have resulted in ecocide, the mass murdering of non-humans to produce a profit. The ecocide underpins a class war where a portion of the human species takes up more space, appropriates more resources, produces and consumes more commodities, and creates more waste. The class war translates to most as poor and precarious working conditions, urban slums, debt bondage, or if you are not so "lucky," as death from above by drone warfare, forced migration, polluted or hostile habitats, animal concentration camps, death, and extinction.[17]

4 Degrowth: Pushing beyond Business as Usual

With each passing moment, the capitalist mode of production (and our lives and societal arrangements that are tied to it) is more unsustainable and unequal. Surely, a paradox of our time is that the concept of sustainability has been coined and advocated in this very same context. It is peculiar that we, the ecological over-shooters in the Global North, cannot stop speaking about sustainability, but at the same time, we do not know what it is, what it looks like, or how to do it in practice. Maybe this is because there is very little cultural, lingual, habitual, and bodily understanding left in the modern and the industrial way of living that would inform us. Perhaps our grandparents or their parents possessed some knowledge of some of the practices, but the boomers or Generations X, Y, or Z do not seem to have a clue. Instead of "going back to nature," as Jean-Jacques Rousseau and many others since his days have suggested, modern folks have tried to cut down their ties to subsistence as the city has "enslaved"

17 T. Ruuska, P. Heikkurinen, T. LeVasseur and A. Gare,"Redefining violence for the Anthropocene: From ecocide to ecological civilization," Cosmos and History: The Journal of Natural and Social Philosophy, 20(2), 56–76 (2024).

the countryside.[18] Like Illich's suggested in *Shadow Work* (1981, 139), it has been a 500-year war against communal subsistence economies. This war still goes on in the Global South with agrobusiness, extractivism, bottom-of-the-pyramid economics, micro credits, and the Western educational curriculum.

Although it has relatively long roots, capitalism (especially in its industrial form) seems to be the antithesis to something that would last a very long time. Instead, it is short-sighted and incapable of rational action, as Marx already declared.[19] Does *sustainability* not mean something that would last for a long time? If it does, should sustainability be measured not in centuries but in thousands of years? Due to the increasingly destructive nature of global industrial capitalism, it is difficult to perceive how it could last for many decades longer, especially when considering the looming peaks in key resources, such as oil, rare earth metals, and various other minerals.[20] And if, up to that point, the capitalist business-as-usual activities are advanced, the meltdown of the global industrial economy will be something else entirely than sustainability. Based on this, we could argue that capitalism is heading towards a collapse.[21] If so, this collapse is likely to be a chaotic event if not prepared for or mitigated beforehand. The members of the degrowth movement see this clearly, and for two decades they have suggested radical reforms aiming to move beyond the economics and politics of perpetual economic growth.[22] Lately, they have also posed explicit demands for organizing beyond capitalism.[23]

In *The Entropy Law and the Economic Process*, published in 1971, Nicholas Georgescu-Roegen argued that economic production should be considered together with biological processes and evolution, and as such, it is subject to the laws of thermodynamics. From an ecological economics perspective, industrial production and expansion accelerate material transformation from low entropy to high entropy, which at some point acts as a fundamental hurdle

---

18 See, e.g., W.E. Rees "Cities, energy and the uncertain future of urban civilization," *Oxford Development Studies*, 51(1), 11–17, (2023).

19 K. Marx, *Capital vol. 3* (Penguin, 1981 [1894]), p. 754; see also J. B. Foster, *Ecological Revolution* (Monthly Review Press, 2009), p. 181.

20 R. Heinberg, *Peak Everything* (New Society Publishers, 2010); J. Bendell, *Breaking Together* (Good Works, 2023); Pineault (2023).

21 Bendell (2023).

22 E.g., G. Kallis, S. Paulson, G. D'Alisa and F. Demaria, *The Case for Degrowth* (Polity, 2020); J. Hickel, *Less is More* (Penguin Random House 2021). T. Parrique, *The political economy of degrowth* (Stockholm Universiteit, 2019). A. Toikka et al. "Degrowth and postgrowth: A systematic literature review of growth-critical science," *Ecological Economics*, 240, 108830, 1–12 (2026).

23 E.g., M. Schmeltzer, A. Vetter and A. Vansintjan, *The Future is Degrowth* (Verso, 2022); N. Barlow, L. Regen, N. Cadiou, *Degrowth & Strategy* (Mayfly Books, 2022).

to economic growth. As Georgescu-Roegen wrote,[24] "The economic process is entropic: it neither creates nor consumes matter or energy, but only transforms low into high entropy." Based on his work and that of many others in the field of ecological economics and other fields of heterodox economics, we can claim that the aim to perpetually grow the size of the economy, especially at a compound rate, eventually contradicts the bio-physical processes of the Earth, and thus, it is not able to sustain itself.

The influence of ecological economics and Georgescu-Roegen's groundwork are significant to degrowth debates and scholarship.[25] Degrowth, as field and a movement, has two main ambitions: (1) to stop the growth of matter-energy throughput globally, and eventually degrow it to a sustainable level, primarily by shrinking the size of the overproducing and overconsuming economies; and (2) create institutions, relationships, and communities that do well in a society that is not based on perpetual economic growth.[26] Barlow et al. (2022, 11) argue that degrowth should be treated as an alternative vision of societies that would be built around ecological sustainability and social justice. As a term, *degrowth* is increasingly utilized by activists and scholars to criticize the hegemony of growth and to offer a counter vision to modern growth-dependent societies. According to the members of the degrowth movement, contemporary societies would have to be radically reorganized to reduce the use of energy and resources so that they reach sustainable levels of production and consumption.[27] The transformation needed in industrialized countries, especially in the Global North, would in all probability also lead to the decrease of overall material standard of living.

In other words, the idea is to slow down and reduce consumption and production in an organized and planned fashion while minimizing the harm to humans and non-humans. From this point, one is likely to realize that degrowth seems to be the opposite to capitalism, where the idea is to accelerate and accumulate. However, the slowdown involved in degrowth does not mean perpetual depression, and thus, it is not reversed capitalism but something else entirely. It does not entail forced deprivation but is rather an aspiration to secure enough for everyone in a post-capitalist and post-growth context.[28] In addition to being a critique and an activist movement, it is also utopian

24 N. Georgescu-Roegen, *The Entropy Law and the Economic Process* (Harvard University Press, 1971), p. 281.

25 E.g., G. Kallis, *Degrowth* (Agenda Publishing, 2018); Parrique (2019); Kallis et al. (2020); Hickel (2021); Schmelzer et al. (2022); Barlow et al. (2022).

26 Kallis et al. (2020), p. 18.

27 Schmelzer et al. (2022), p. 3.

28 Kallis et al. (2020), p. 18–19.

project that calls on people to think, act, plan, and strategize for alternative futures. However, degrowth scholars and activists do not claim to possess a unitary plan or a road map that would fit everywhere, all the time. It is instead a diverse network of various initiatives and debates about what degrowth is and how it should be put into action in different contexts.

The degrowth movement is based on the principles of autonomy, solidarity, direct democracy, and bottom-up organizing that seeks to influence politics and economies in a just manner.[29] Although the concept is European, as are most of the people who currently form the movement, degrowth is not blind to the historical inequalities of (neo)colonialism and imperialism, and therefore, the movement demands that the affluent economies of the Global North withdraw from the operations that cause social and ecological burdens to the nations in the Global South.[30] In many ways the origins of degrowth are understandable since most of the overproducing and overconsuming economies are found in the Global North, which means that the movement has been born at the heart of the problem. Meanwhile, degrowth also connects to a pluriverse of ideas around the world that seek well-being that goes beyond capitalism and (neo)colonial development policies.[31] It tries to picture and build up societies and spaces where one's well-being does not mean living at the expense of others.[32]

To reach this fundamental political and economic reorganization of industrial, growth-based societies, it is necessary to overcome the multiple societal and structural growth dependencies of capitalism. These vary from industrial and non-renewable resource-based infrastructures that are powered by fossil fuels to social constructs concerning technology and economic growth. The hegemony of capital and growth sits hard on most people inhabiting industrialized societies, and it is no wonder that degrowth has many critics who often state (especially in a time when the capitalist organization is in crisis) "see what happens when the economy does not grow."[33] To be sure, economic downturn in a capitalist economy is a catastrophe for many, and it is understandable that people connect degrowth to uncertainty or even full-fledged collapse. But

29 E.g., V. Asara, E. Profumi, G. Kallis, "Degrowth, Democracy and Autonomy," *Environmental Values* 22, n. 2 (2013).

30 See e.g., Barlow et al. (2022), p. 11.

31 E.g., A. Kothari, A. Salleh, A. Escobar, F. Demaria and A. Acosta, *Pluriverse* (Tulika Books, 2019).

32 See U. Brand, B. Muraca, E. Pineault et al. "From Planetary to Social Boundaries," *Sustainability: Science, Practice and Policy*, 17, n. 1 (2021).

33 For a comprehensive collection of critiques towards degrowth, see Parrique (2019), Chapter 7.

this uncertainty is exactly the thing that degrowth policies, seeking radical reforms, try to avoid. Degrowth does not mean recession (or politics based on recession) because recession is unintentional, whereas degrowth seeks to be planned and intentional. Schmeltzer et al. (2022, 20) noted that: "recessions make inequality worse, degrowth seeks to reduce it; recessions typically lead to cuts in public services while degrowth is about de-commodifying essential goods and services; recessions often cause bold policies for sustainability to be abandoned for the sake of restarting growth, while degrowth is explicitly for a rapid and decisive transformation."

Overall, degrowth activists and researchers pursue eco-societal transformations that would take us beyond capitalism by reorganizing current societal structures and power relations. Although for some, *degrowth* is a missile word and a radical statement in a time when everything still revolves around economic growth, it is for many a bundle of arguments used to challenge the hegemony of capital and also a call to start acting, strategizing, and planning for the future beyond growth. While it may be true that the critiques and principles of degrowth are finding a larger audience, at the same time, it is true that many of those demands are still marginal, abstract, and intangible. Although countless things fall under the degrowth umbrella nowadays, a common idea of where to degrow from the overshoot seems to be missing. It is in regard to this that degrowth's sister concept, sufficiency, may prove to be helpful.

5 Sufficiency in Affluence and Technology

As an idea, *sufficiency* is a straightforward one: it refers to having *enough*, not too little or too much but just the right amount required for a certain task, need, or policy, or for organizing.[34] Although the contemporary capitalist societies operate as if nothing is enough, the deepening ecological crisis has arguably turned the tables to the point where societies must recognize their "self-generated predicament and accept that enough is enough."[35] In a recent literature analysis on the concept, Jungell-Michelsson and Heikkurinen (2022, 6) have produced the following definition of *sufficiency*:

> Sufficiency is a transdisciplinary concept about "enoughness" of human doings in relation to ecosystems—an end in itself and a means for sustainable consumption and production comprising three main premises,

34 E.g., T. Princen, "Principles for Sustainability," *Global Environmental Politics* 3, no. 1 (2003); T. Princen, *The Logic of Sufficiency* (MIT Press, 2005).

35 Princen (2003), p. 43.

> namely the complementarity of capital (from ecological economics), social metabolism (from political ecology), and altruism towards human and non-human beings (from ecological philosophy).

As stated in their analysis, in the field of ecological economics the discussion on sufficiency has revolved around the critique of using efficiency as a guiding principle of sustainability and the economy. In contrast, in the field of political ecology sufficiency has been treated in a more normative fashion, for instance, "as a field of action instead of referring to individual decisions and lifestyles as today."[36] Accordingly, and similarly to debates on degrowth, a slowdown and a qualitative change in metabolism are necessary to repair the produced metabolic rifts that are due to exploitative production processes and the overall workings of capitalism.[37] Moreover, in the field of environmental philosophy, sufficiency has been discussed in relation to a shift to non-material values that facilitate well-being and a more just distribution of resources between humans and non-humans.

As Princen (2003, 45) noted, sufficiency as a concept, notion, or an end does not necessarily carry anything new: "Historically, societies have developed related notions—moderation, thrift, frugality, prudence, temperance, reverence—all to restrict the otherwise human tendency to want more of a good thing. The risks were to social cohesion, self-defense, and survival." Similarly to degrowth, sufficiency may carry the seeds of change because it begs us—the overshooters—to ask, "How much is enough?" or "What and how much is needed?" It could be argued that in each cultural, societal, and communal context, humans only need enough to provide them with a flourishing, just, and fulfilling life, and they only need that amount since humans need to consider both the planet that is full of their kind and the conditions required for the continuation of diverse life on planet Earth.

In the remaining two sub-sections, and before the conclusion, affluence and technology are discussed from the point of view of degrowth debates in order to find potential paths towards sufficiency.

5.1 *Local and Communal Affluence*

Industrial capitalism and its underlying processes have created metabolic rifts, resulting in the degradation of the soil (the soil's nutrients being lost to the

36 S. Lorek and J. Spangenberg, "Energy sufficiency through social innovation in housing," *Energy Policy* 126 (2019), p. 293.

37 See A. Salleh, "From Metabolic Rift to "Metabolic Value"," *Organization & Environment* 23, no. 2 (2010); see also Foster (2000) and Saito (2023).

rivers, streams, seas, sewage systems, and garbage dumps).[38] Recently, Saito[39] has made an important contribution to Foster's and Moore's metabolic rift/shift conceptions[40] by including three dimensions of metabolic rift (material, spatial, and temporal dimensions) and three metabolic shifts (technological, spatial, and temporal shifts) to highlight the disruptive and destructive workings of capital(ism) in nature. To supplement this important conception, I would like to argue that there also exists *a cultural rift,* and *a cultural shift. Cultural rift* signifies the gradual erosion and dismantling of the local and communal subsistence economies of the countryside/hinterlands/wilderness during the era of industrialization and capitalist globalization.[41] *Cultural shift* denotes then to a transformation from land-based livelihoods and communal ways of living to the life of the urban and to individualist mass consumerism upholding a global restructuring of labor and individualist identities and identity politics.

The division of labor has been deepening since various agricultural innovations and since the restructuring of property and class relations, first in rural England and then elsewhere, in the fifteenth and sixteenth centuries.[42] Land enclosures and competition between farmers led to rising productivity, which created surplus laborers who would flow to the cities and eventually form the industrial proletariat. More broadly, technologization and commodification have resulted in the automatization, industrialization, and capitalization of the rural, and at the same time, they have accelerated urbanization, which has had eroding effects on communal and local subsistence economies. This is truly a cultural turn, a turn from predominantly small and communal village and town economies with direct ties to the surrounding environment and people to the life of the metropole, which appropriates people, resources, and non-human beings to reproduce the processes of industrial and global capitalism.[43]

This cultural turn also signifies a transformation from communal identities to individualism[44] and from place-based relationships to rootlessness and urban identity politics. In some places, mostly in the Global South, local and communal economies may still thrive, but they are constantly at risk of being wiped out by agrobusiness, development projects, and fossil-fueled urbanization. These prevailing worldwide land-based communities largely secure their

38 See Foster (2000).

39 Saito (2023), pp. 23–34; see also Ruuska (2021).

40 Foster (2000); Moore (2015).

41 T. Ruuska, "A Path to Food Self-Provisioning and Experiences from Learning New Skills," *The Qualitative Report* 29, no. 2. (2024).

42 See E. M. Wood, *The Origin of Capitalism* (Verso, 2002), p. 100; Moore (2015), pp. 242–243.

43 See Saito (2023), Chapter 1.; Rees (2023).

44 E.g., C. Taylor, *The Ethics of Authenticity* (Harvard University Press, 2018).

basic needs by themselves with relatively low and slow technology, and they are also embodied and identified by a place. The contrast to the hyper-developed mass consumerist metropoles is vast. As we already know, these urban industrial centers have taken a heavy toll on the rural ecocultures: global supply chains and trade policies have furthered the short-sighted agenda of industrial and commercial centers at the expense of the short- and long-term viability of the rural.[45]

Considering that the most academics and many activists nowadays reside in urban areas, as do most humans on Earth, it is no wonder that sustainability initiatives and movements are often city-based. The same goes for the degrowth movement, which up to this day has not sufficiently paid attention to the rural political economy, to the so-called agrarian question,[46] or to the progressive sustainability prospects of the rural.[47] At the same time, many degrowth activists and scholars seem to assume that cities can become sustainable. However, are cities not places where too many people live compared with the bioregion's carrying capacity? As Smaje wrote:

> I think it's worth asking whether your locality can feed its population. [...] For many people the answer will be no, because you live in a city where population density is too high to produce enough food, although some places could still be identified for meeting a proportion of the city's needs. For people whose localities can't meet their needs, the question is which other localities might fill the breach—and why.[48]

The unsustainability of large cities calls for a fundamental reexamination of how people should reorganize their lives in over-affluent societies in planned and controlled degrowth transitions. It is difficult to perceive how sufficiency lifestyles could come about without localizing and deurbanizing the overshooting industrial economies and societies. Meanwhile, life in the rural cannot be based on technologization, fossil fuels, and industrial agriculture, it must be based on qualitatively different relationships and usage of land. Like Shiva (2008) wrote, *localization* is a food-security and food-justice imperative,

---

45 E.g., V. Shiva, *Soil not Oil* (Zed Books, 2008); S. Böhm, Z. P. Bharucha and J. Pretty, *Ecocultures* (Routledge, 2015); T. LeVasseur, T. Ruuska and P. Heikkurinen, "Imagining a Prosperous Periphery for the Rural in 2050 and beyond," In Filho et al., eds., *Handbook of Sustainability Science in the Future* (Springer, 2022).

46 See J-F. Gerber, "Degrowth and critical agrarian studies," *Journal of Peasant Studies* 47, no. 2 (2020).

47 E.g., Shiva (2008); C. Smaje, *A Small Farm Future* (Chelsea Green Publishing, 2020).

48 Smaje (2020), p. 151.

too. As she also noted (ibid., 126), collaborative and local peasant economies can "provide the care and attention required to facilitate nature's self-organization" and the conditions for just and sufficient well-being if these communities empower their representatives in an equal fashion.

There is indeed a need to preserve and rebuild local and communal economies on the basis of degrowth, eco-sufficiency, and social justice.[49] As Salleh (2009, 414) argued, "localized metabolic provisioning is eco-sufficient because it does not externalize costs through debt or entropy, or by passing them off to future generations."[50] The equal, diverse, and affluent local communities could be built around the three interlinking aims that she (ibid., 417) highlighted: (1) protecting the material/energetic interdependency of species, sex/genders, and generations; (2) protecting metabolic value and the sovereignty of common livelihoods; and (3) protecting and cherishing the difference and diversity of various, economic, social, and cultural practices. Together these aims have the potential to produce qualitatively different local livelihoods and ways of life that would not be based on increasing the material standard of living or affluence at the expense of others.

5.2 *Modern Technology versus "Slower and Lower" Technology*

In his inspiring account on revolutions, Enzo Traverso touched on the two contradictory streams of Marx's and Marxist thought on history and technology. As Traverso (2021, 52–53) noted, socialism, for many, has meant the liberation *from* labor and not liberation *through* labor, that is, it has meant liberation from hard work that is now conducted by machines. The idea is grounded in a particular socialist utopia where "intellectual and physical faculties could be devoted to the accomplishment of creative tasks rather than to the satisfaction of primary needs."[51] It is a utopia according to which humans are emancipated from muscular labor and material constraints, but concurrently, as Traverso (ibid.) stated, there is "a dangerous idealization of technology that announces the controversial relationship between socialism and ecology in the twentieth century."

As Traverso also wrote, Marx's impressive body of work is driven forward by an unresolved tension between two contradictory tendencies. On the one hand, a positivistic and deterministic tendency to see history and technology as progressive—for instance, to portray communism as the inevitable

49 See Saito (2023).

50 A. Salleh, "From eco-sufficiency to global justice," In A. Salleh, ed., *Eco-Sufficiency and Global Justice* (Pluto Press, 2009).

51 Traverso (2021), p. 53.

successor of capitalism—or a tendency to accuse capitalism for the failures of technology. On the other hand, Marx surely had a dialectical and materialist conception of history, according to which history is seen as an open and non-teleological process, without any predetermined direction, although its direction can be influenced and shaped by human agency. In the latter one of the two tendencies, the development of productive forces and technology is also subject to a negative dialectic where the development of technology could lead to further oppression of the working class and to the destruction of natural habitats.[52] This tension between determinism and constructivism, as Traverso claimed, was not reconciled in Marx's work, nor has it been sufficiently resolved by his followers.[53] Although Marxists started to abandon the conception that considers technology as something neutral after the two catastrophic world wars, the dialectic of technology, which simultaneously seems to uphold the promise of liberation and the curse of mass murder, remains underexplored and undertheorized.[54]

The inability to recognize the dialectic and perils of technology is, to some extent, also prevalent in the degrowth movement, whose relation to technology remains ambivalent.[55] Similarly to many Marxists, the blind spots of the proponents of the degrowth movement seem to lie both in the ecological footprint of technology and in the assumption that technology, and its development, can become just and equal in the future. If this is so, the general take of the degrowth movement on technology is ambivalent in much the same fashion as was the take of the bulk of the twentieth century Marxists.

Going back to ecologically unequal exchange, Hornborg (1998; 2009; 2016) has sought to demonstrate that the question of technology and its development cannot be solved fairly from a socio-economic or ecological standpoint.[56] In brief, this is because technology cannot be decoupled or detached from global power relations, inequality, or local environmental hazards. He claims that even though technology is represented as a silver bullet solution to contemporary eco-social maladies, the development of technology and its utilization have produced the opposite result: technologization has increased global socio-economic disparities and environmental conflicts and problems. For

52 Traverso (2021); See also Saito (2023).

53 See Saito 2023, for Marx's post *Capital vol. 1* turn away from productivism, eurocentrism and ethnocentrism.

54 See Traverso (2021), pp. 55–56; H. Marcuse, *One-Dimensional Man* (Routledge, 2002 [1964]); see also Ruuska (2021).

55 E.g., C. Kershner, P. Wäcter. L. Nierling and M.H. Ehlers, "Degrowth and Technology," *Journal of Cleaner Production* 197, part 2 (2018).

56 See also Dorninger et al. (2021).

well over twenty years, Hornborg has argued that the reasons for this are the ecologically unequal exchange of the global capitalist economy and the way in which people tend to perceive technology.

It is rather easy to picture how many Marxists, and perhaps also some degrowth researchers and activists, would respond to Hornborg's claims. They would likely argue that the ecologically unequal exchange results from the capitalist relations of production and oppression, and/or from the global power imbalance, and not from technology per se. Surely, capitalism and the underlying power imbalances are part of any critical reasoning related to the problems of the status quo, but it should also be clear that there is no societal structure that would automatically do away with the ills of modern, resource-intensive technology. This is because there is no relation of production that downplays the ecological impacts of the extractivism that is related to solar panels, wind turbines, car batteries, computers, nuclear power plants, or industrial infrastructure in general, above all because natural resources and their extraction are always local issues, too. Those resources are extracted from someone's backyard or home at the expense of that person and for the benefit of someone that lives beyond the premises. Therefore, the use and manufacturing of modern, resource-intensive technology are always ethically and matter-energetically contested questions.

As Heikkurinen (2018) has argued, the development, manufacturing, and use of technology always requires the transformation of nature. As we are dealing with an increasing amount of resource-intensive modern technology, this transformation, or the destruction of natural habitats, is at an unprecedented level. It may be that communist, degrowth, or degrowth communist technology would be developed differently and for better purposes than profit-making, mass surveillance, and warfare, but if that technology retains its resource-intensive form, it will not do away with the negative ecological and social consequences.[57] The only sure way to avoid the undesired and uneven consequences of modern technology is to minimize its use and development, and to push for lower and slower technology that is mainly powered by biomass or muscle power and has a long lifecycle. For those residing in local and communal economies, this is not a significant change of things, but for those in the metropoles, it might seem like the end of things, and in a way, it is.

Even though this seems impossible and ridiculous for the metropolitans and all the billions of people hooked on their devices, we should remember

57 See P. Heikkurinen and T. Ruuska, eds., *Sustainability beyond Technology* (Oxford University Press, 2021).

once again that modern technology is by no means sustainable. It is addictive, unethical, unequal, and ecologically disastrous. Although it has temporarily freed some portions of humans from muscular labor, it has done so at the expense of the living conditions of countless other beings on this planet. In contrast, the push for sufficient technology is again a local question that should be solved by local means to an end. Slow and low technology does not mean that we should turn our backs on the rest of the world or move into a cave, but it means that we do not have to develop, manufacture, and use technology at the expense of others.

6 Conclusion

The contemporary world economy is driven by perpetual economic growth, competition, and the accumulation of capital. This system relies on the exploitation of others. These others are our fellow humans, working in sweat shops and on industrial assembly lines, as well as dispossessed peasants and small holders, mainly in the Global South, and our fellow non-human beings all over the biosphere.

Instead of more growth and affluence, there is a call for the redistribution of wealth and resources, and the degrowth of the overshooting economies so that they move towards sufficiency and sustainability. In this sense, degrowth could be considered as journeys or alternative paths away from the human-induced ecological crisis and inequality, and toward sufficient dwelling and livelihoods as destinations where all people have enough.

To move towards sufficiency in the times of great uncertainty and distress, we need to be able to discuss and practice local and communal affluence, and foster lower and slower technology. For the degrowth movement, this means questioning urban dominance and urbanization-favoring sustainability initiatives, as well as pushing for local/communal livelihoods and resilience.

Acknowledgements

This work was supported by funding from the Research Council of Finland's project "The *Skills of Self-provisioning in Rural Communities*", grant number 343277, and the Kone Foundation's project "The *Underdogs of Just Transition*" (*Oikeudenmukaisen siirtymän altavastaajat*), grant number 202302807.

References

Arboleda, M. (2020). *Planetary Mine: Territories of Extraction under Late Capitalism*. Verso.

Asara, V., Profumi, E., Kallis, G. (2013). "Degrowth, Democracy and Autonomy," *Environmental Values*, 22(2): 217–239.

Barlow, N., Regen, L., Cadiou, N. et al. (2022). *Degrowth & Strategy: How to bring about social-ecological transformation*. Mayfly Books.

Bendell, J. (2023). *Breaking Together: A freedom-loving response to collapse*. Good Works.

Brand, U., Muraca, B., Pineault, E. et al. (2021). "From Planetary to Social Boundaries: An Argument for Collectively Defined Self-Limitation," *Sustainability: Science, Practice and Policy*, 17(1): 264–291.

Böhm, S., Bharucha, Z. P. and Pretty, J. (2015). *Ecocultures: Blueprints for Sustainable Communities*. Routledge.

Dorninger C., Hornborg, A., Abson, D., et al. (2021). "Global patterns of ecologically unequal exchange: Implications for sustainability in the 21st century," *Ecological Economics*, 179: 1–14.

Dunlap, A. and Jakobsen, J. (2020). *The Violent Technologies of Extraction: Political ecology, critical agrarian studies and the capitalist worldeater*. Springer.

Foster, J. B. (2000). *Marx's Ecology: Materialism and Nature*. Monthly Review Press.

Foster, J. B. (2009). *Ecological Revolution: Making Peace with the Planet*. Monthly Review Press.

Georgescu-Roegen, N. (1971). *The Entropy Law and the Economic Process*. Harvard University Press.

Gerber, J.-F. (2020). "Degrowth and critical agrarian studies," *Journal of Peasant Studies*, 47(2): 235–264.

Givens, J. G., Huang, X. and Jorgenson, A. (2019). "Ecologically unequal exchange: A theory of global environmental injustice," *Sociology Compass*, 13(5): 1–15.

Heikkurinen, P. (2018). "Degrowth by means of technology? A treatise for an ethos of releasement," *Journal of Cleaner Production*, 197(2): 1654–1665.

Heikkurinen P., and Ruuska T. (eds.) (2021). *Sustainability Beyond Technology: Philosophy, Critique, and Implications for Human Organization*. Oxford University Press.

Heinberg, R. (2010). *Peak Everything: Waking Up to the Century of Declines*. New Society Publishers.

Hickel, J. (2021). *Less is More: How Degrowth Will Save the World*. Penguin Random House.

Hornborg, A. (1998). "Towards an ecological theory of unequal exchange: articulating world-system theory and ecological economics," *Ecological Economics*, 25(1): 127–136.

Hornborg, A. (2009). "Zero-Sum World: Challenges in Conceptualizing Environmental Load Displacement and Ecologically Unequal Exchange in the World-System," *International Journal of Comparative Sociology*, 50(3–4): 237–262.

Hornborg, A. (2016). *Global Magic: Technologies of Appropriation from Ancient Rome to Wall Street.* Palgrave Macmillan.

Illich, I., (1981). *Shadow Work.* Marion Boyars Publishers.

Jungell-Michelsson, J. and Heikkurinen, P. (2022). "Sufficiency: A systematic literature review," *Ecological Economics*, 195: 1–13.

Kallis, G. (2018). *Degrowth.* Agenda Publishing.

Kallis, G., Paulson, S., D'Alisa, G. and Demaria, F. (2020). *The Case for Degrowth.* Polity.

Kerschner, C., Wäcter, P., Nierling, L. and Elhers, M.-H. (2018). "Degrowth and Technology: Towards feasible, viable, appropriate and convivial imaginaries," *Journal of Cleaner Production*, 197, part 2: 1619–1636.

Kothari, A., Salleh, A., Escobar, A., Demaria, F. and Acosta, A. (2019). *Pluriverse: A Post-Development Dictionary.* Tulika Books.

LeVasseur, T., Ruuska, T. and Heikkurinen, P. (2022)."Imagining a Prosperous Periphery for the Rural in 2050 and Beyond," in *Handbook of Sustainability Science in the Future: Policies, Technologies and Education by 2050*, Filho, W. L. et al. (eds.). Springer Nature.

Lorek, S., and Spangenberg, J. (2019). "Energy sufficiency through social innovation in housing," *Energy Policy*, 126: 287–294.

Marcuse, H. (2002 [1964]). *One-Dimensional Man.* Routledge.

Marx, K. (1976 [1867]). *Capital: A Critique of Political Economy, Volume 1: The Process of Capitalist Production.* Translated Fowkes, B. Penguin.

Marx, K. (1981 [1894]). *Capital: A Critique of Political Economy, Volume 3: The Process of Capitalist Production as a Whole.* Translated by Fernbach, D., Penguin.

Moore, J. W. (2015). *Capitalism in the Web of Life: Ecology and the Accumulation of Capital.* Verso.

Moore, J. W. (2017). "Metabolic rift or metabolic shift? Dialectics, nature, and the world-historical method," *Theory and Society*, 46: 285–316.

Parrique, T. (2019). *The political economy of degrowth. Economics and Finance.* Université Clermont Auvergne; Stockholm Universiteit.

Pineault, E. (2023). *A Social Ecology of Capital.* Pluto Press.

Princen, T. (2003). "Principles for Sustainability: From Cooperation and Efficiency to Sufficiency," *Global Environmental Politics*, 3(1): 33–50.

Princen, T. (2005). *The Logic of Sufficiency.* MIT Press.

Rees, W. E., 2023. Cities, energy and the uncertain future of urban civilization. Oxford Development Studies, 51(1), 11–17.

Ruuska, T. (2021). "Conditions for Alienation: Technological Development and Capital Accumulation," in *Sustainability Beyond Technology: Philosophy, Critique, and*

Implications for Human Organization, Heikkurinen, P. and Ruuska, T. (eds.). Oxford University Press.

Ruuska, T. (2024). "A Path to Food Self-Provisioning and Experiences from Learning New Skills: An Autoethnographic Depiction," *The Qualitative Report*, 29(2): 484–501.

Ruuska, T., Heikkurinen, P., LeVasseur, T., and Gare, A. (2024)."Redefining violence for the Anthropocene: From ecocide to ecological civilization," Cosmos and History: The Journal of Natural and Social Philosophy, 20(2), 56–76.

Saito, K. (2023). *Marx in the Anthropocene: Towards the Idea of Degrowth Communism*. Cambridge University Press.

Salleh, A. (2009). "From Eco-sufficiency to Global Justice," in *Eco-sufficiency and Global Justice: Women write political ecology*, Salleh A. (ed.). Pluto Press.

Salleh, A. (2010). "From Metabolic Rift to 'Metabolic Value': Reflections in Environmental Sociology and the Alternative Globalization Movement," *Organization & Environment*, 23(2): 205–219.

Schmeltzer, M., Vetter, A. and Vansintjan, A. (2022). *The Future is Degrowth: A Guide to a World beyond Capitalism*. Verso.

Shiva, V. (2008). *Soil Not Oil: Environmental Justice in an Age of Climate Crisis*. Zed Books.

Smaje, C. (2020). *A Small Farm Future: Making the Case for a Society Built Around Local Economies, Self-provisioning, Agricultural Diversity, and a Shared Earth*. Chelsea Green Publishing.

Steffen, W., Broadgate W., Deutsch L., et al. (2015). "The trajectory of the Anthropocene: The great acceleration," *Anthropocene Review*, 2(1): 81–98.

Taylor, C. (2018). *The Ethics of Authenticity*. Harvard University Press.

Toikka, A., Ruuska, T., Wiman, L. and Heikkurinen, P. (2026). "Degrowth and post-growth: A systematic literature review of growth-critical science," *Ecological Economics*, 240, 108830, 1–12.

Traverso, E. (2021). *Revolution: An Intellectual History*. Verso.

Wood, M. E., (2002). *The Origin of Capitalism: A Longer View*. Verso.

CHAPTER 6

A Business of Deep Transformations Is a Business of Sufficiency

Iana Nesterova and Jessica Jungell-Michelsson

1 Introduction

In the face of unfolding and intensifying ecological degradation, scholars and practitioners contemplate what a genuinely sustainable society may look like, what it may entail, which social entities and social structures will be features of its landscape, and how to get there.[1] Amongst such scholars are the scholars of degrowth. Degrowth is a broad and diverse research field loosely united by the critique of pursuit of economic growth[2] and a desire for a different socio-economic system where positive outcomes for humans and nature are not linked to economic growth.[3] In its contemplation of a different, more sustainable socio-economic system, degrowth tends to pay particular attention to the affluence (A) and technology (T) elements of the IPAT equation. Affluence[4] is argued to acquire a different meaning altogether: affluent societies are ones where satisfaction of everyone's basic needs co-exists with non-material growth and utmost respect for nature and its limits.[5] Technology, in turn, is not seen as the solution to ecological degradation, but rather as one of multiple means which needs to be critically evaluated.[6]

Considering the breadth of the research field of degrowth as well as its diversity, we find it essential to emphasise the definition that we use. We adopt the definition of degrowth that is taking shape in the Nordics where our authorship team is based. We acknowledge, alongside with other degrowthers, the need to deviate from the pursuit of economic growth for its own sake. However, we highlight growth in the non-material and in human moral agency, i.e.,

1 Bonnedahn and Heikkurinen 2019; Paech 2012.

2 Economic growth is measured in terms of increase of GDP (Gross Domestic Product). GDP is the total value of goods and services produced in an economy.

3 Buch-Hansen et al. 2024.

4 Affluence in mainstream terms can be reflected in GDP per capital (GDP divided by a country's population).

5 Buch-Hansen et al. 2024.

6 Heikkurinen and Ruuska 2021.

 | DOI:10.1163/9789004746404_007

growth in care, solidarity, empathy, creativity, generosity, and connectedness with the self, others, and nature.[7] We likewise highlight the need for growth in certain industries which satisfy human needs while manifesting gentleness and care towards nature.[8] Thus, we do not equate degrowth with reduction in all economic and business activities. The industries where growth is desirable include, for instance, organic agriculture, alternative housing, permaculture, small scale and decentralised energy production, service providers such as zero waste stores. We also encourage growth in consumption of those humans whose basic and genuine human needs, such as the needs for food and water, shelter, warmth, medicine, and education have not been sufficiently met. Hence, despite the negative connotations associated with the word degrowth[9] the message we encourage in our approach is that of growth in harmony, humanness, and flourishing. However, such growth necessarily needs to co-exist with the considerations of what is enough or sufficient in material terms.

In recent times, degrowth scholarship has contributed substantially to both creating images of a possible genuinely sustainable society and to our understanding of various pathways towards such society. However, degrowth is not a uniform field. There remain many disagreements as to what a degrowth society will entail. This includes, for instance, what degrowth would mean in terms of our political system, whether it would be anarchist, socialist, some combination of these or a different system altogether. Likewise, it is still unclear how a degrowth society would come about.[10] Despite these disagreements and lack of clarity, what is becoming increasingly clear is that our current materially wealthy and unsustainable (overshooting) societies need to undergo deep transformations.[11] Deep transformation can be contrasted with techno-optimism which postulates that the issues of ecological degradation will be solved via technological advancement and intensification in the use of "green" technologies. The deep transformations discourse contends that technological innovations, including the (supposedly) green ones, will not resolve ecological degradation. Technology on its own will not be able to ensure co-existence between humanity and nature.[12] Neither is decoupling economic growth from resource use, assumed to be possible by techno-optimists, in fact feasible.[13]

7 Buch-Hansen 2021; Nesterova 2022a.
8 Buch-Hansen and Nesterova 2023.
9 Drews and Antal 2016.
10 Buch-Hansen 2014.
11 Buch-Hansen and Nesterova 2021; Nesterova 2022b.
12 Heikkurinen 2018.
13 Paech 2017; Trainer 2022.

The deep transformations discourse within degrowth is not opposed to technology. Rather, it avoids seeing technology as neutral and adopts a more philosophical approach to change which is seen as complex and non-linear. Such change requires nurturing empathy, solidarity, gentleness, and care. It also requires sufficiency orientation (e.g., in material possessions, speed of innovation, growth) and place-sensitivity and needs to happen on all levels of societal existence, from individual psyches to social systems.[14] Deep transformations concern everyone who dwells in overshooting societies. Such transformations entail not only *transformation* of societal structures which are detrimental to the co-existence between humans and nature. Deep transformations also entail *reproduction* of those structures which are supportive of co-existence. In other words, it is not only currently unsustainable structures that need to change. Sustainable structures need to be intentionally and actively reproduced, and the agents currently involved in sustainable structures need to share their pathways to inspire change in various other domains of societal existence.

Undoubtedly, important actors in deep transformations will be business organisations. To exclude business organisations from discussions and the practice of deep transformations is not a suitable option. They are essential to many core activities in current societies. Even though there are alternative producing and service providing organisations (e.g., repair cafes, peer production organisations) which may be more compatible with degrowth due to their community based and explicitly political (e.g., activist) and non- (or even anti- or post-) capitalist approaches, mainstream business organisations[15] still make up a large part of the economic activities in current societies and therefore cannot and should not be ignored. Moreover, the fact that a business organisation needs to seek and make a profit in a capitalist system does not necessarily disqualify it from participating in transformation and self-transformation towards an alternative system. Rather, what needs to be pursued is deviation from the centrality of profit-seeking while the system transforms towards one where profit as a pursuit becomes redundant. A business that seeks to deviate from the centrality of profit-seeking and steps on a path of ecologically and socially harmonious existence is referred to as a degrowth business.[16]

In this chapter, we use the terms "degrowth business" and a "business of deep transformations" interchangeably. While the term "degrowth business"

---

14 Buch-Hansen et al. 2024.

15 In the forms of e.g., limited company (Ltd) in the UK, "aktiebolag" (Ab) in Sweden or "osakeyhtiö" (Oy) in Finland.

16 Nesterova 2020.

originally designated a concept of a business for a degrowth society, "business of deep transformations" captures the essence of what a degrowth business entails. Degrowth business needs to be transformative towards societal structures as well as transformative towards itself, thus becoming a business of deep transformations.[17] It is important that degrowth businesses are not seen in binary, "yes and no" terms. Viewing businesses as either degrowth or not degrowth is unhelpful: degrowth business is always a process of (self)learning, making mistakes and collectively as well as individually looking for ways in which businesses can become genuinely sustainable.[18] Degrowth business is a journey through the current capitalist landscapes.[19] This journey is ongoing since the capitalist landscapes are always in becoming.

What deep transformations mean in practice for both business and other producing organisations is not clear-cut. Degrowth business practices[20] may include, among others, production for needs satisfaction, producing durable goods, using natural and renewable materials, using renewable energy, being socially responsible.[21] While there is no lack of proposed degrowth business practices in the literature, there remains a need to identify the underlying key principles from which many or even all the practices will stem. There also remains a need to explicate how certain principles relate to concrete business practices.

In this chapter we suggest that one such key principle is *sufficiency.* Sufficiency has previously been highlighted as a key principle and imperative in post-growth economies.[22] We propose that a business of deep transformations will include multiple elements of sufficiency. In the context of sustainability, sufficiency is usually understood as *enoughness* of human doings in relation to ecosystems—an end in itself and a means for sustainable consumption and production'.[23] Sufficiency is often manifested in individual consumption moderation in such a way that a level of *enough* is pursued rather than *ever more*. Additionally, sufficiency has also been conceptualised in terms of production, implying a focus on quality instead of quantity and a paradigm shift of business logics. In this chapter, we discuss sufficiency from the perspective

17 Nesterova 2022b.

18 Nesterova 2024a.

19 Nesterova 2022c.

20 There is, to a large extent, a substantial overlap between the features of degrowth practices and the features of sufficiency (see e.g., Sandberg 2021 on sufficiency transitions and Buch-Hansen et al. 2024 on degrowth transformations).

21 Hankammer et al. 2021; Nesterova 2021a.

22 Paech 2017.

23 Jungell-Michelsson and Heikkurinen 2022, p. 6.

of business practices and contend that sufficiency should become central to production processes and increasingly complement other currently adopted means for achieving sustainability, such as eco-efficiency. Moreover, following Nesterova[24] we adopt a geographical perspective to enrich our link between a business of deep transformations and sufficiency by discussing the enoughness of a location. Thus, we pursue the following dimensions: practices of sufficiency and the sufficiency of location. Finally, we highlight the social context and the social structures (with a focus on consumption) as fundamental for sufficiency and deep transformations. To supplement our theoretical argument, we provide several illustrative cases which show that sufficiency is manifested differently in different businesses.

2 Business Practices of Sufficiency

To structure our exploration of business practices and relate them to sufficiency, we use the expanded dimensions of the degrowth business framework.[25] The dimensions outlined in Table 6.1 include both nature at large and a business' own location, humans (employees, customers, civil society at large), non-humans, and the self. In what follows we relate each of the dimensions to sufficiency and show how sufficiency can serve as an underlying principle of deep transformations of business and how it can be manifested in business practices.

Nature at large and location. At its core, the call for deep transformations arises from the unfolding ecological degradation and the acute need to address it to sustain life on Earth, both human and non-human. For a long time, ecological degradation has been addressed in business as taking "the environment" into consideration in business operations and seeking win-win solutions (good for business and good for nature). "The environment" appears to us to be an excessively broad term, which is better understood as both nature at large and location where a business is based, i.e., a more intimate environment of a business organisation, including characteristics such as the topography and climate. This brings our attention to the need to be knowledgeable about and responsible towards places and be deeply embedded in them.[26] Such deep embeddedness would complement current business considerations that merely aim for "greener" or more "environmentally friendly" operations. However, a business

24 Nesterova 2022b; Nesterova 2022c; cf. spatial confinement of Paech 2017, p. 483.

25 Nesterova 2020; Nesterova 2022c.

26 Nesterova 2022b, 2022c; Whiteman and Cooper 2011.

is rarely only in one place. Even a locally produced bar of soap (say, in Northern Sweden) can require importing ingredients from various countries: shea butter from Ghana, essential oil from France. Considering nature at large entails shortening of supply chains and thus withdrawing from, or at least minimising, activities in various parts of the world, especially exploitative activities or activities a business cannot effectively monitor and control. At the same time, shortening of supply chains means taking a closer look at local conditions, i.e., focusing on the "sufficiency of location". Sufficiency of location implies localisation of production as well as consumption of both goods and services. It entails localisation of employment in those industries which have traditionally sought cheap workforce elsewhere, such as wild berry picking in Northern Europe. We explore this further in section 3 below.

Human beings. In terms of human beings, businesses can encourage sufficient consumption. This can be done via focusing on consumers' needs rather than wants and omitting artificial creation of ever new demands for existing and new products. Sufficiency would then be manifested in alternative business models such as product-service systems or slower innovation cycles, as well as fundamental changes in sales and marketing strategies.[27] Sufficiency could also be manifested in limiting the size of business in terms of the number of employees. For a degrowth society, it appears that a large number of small, human-scale businesses is better than a small number of large businesses due to power that such businesses tend to acquire and accumulate, as well as due to hierarchies, disempowerment and inequalities that inevitably arise in large and complex organisations.[28] Smaller size of business can encourage a focus on sufficient rather than ever expanding range of products. This would allow the business to monitor, control and enhance the quality of those products rather than direct resources and creative capacity to creating more demand as is the case, for instance, with fast fashion companies. In this case, a business can oversee the production of a product as a whole rather than merely its parts. Moreover, implementing sufficiency into business models could potentially create more jobs: when, for instance, lower technology is used, productivity may fall, and thus more human participation is required. For instance, in small scale organic farming the use of artificial pesticides and herbicides is avoided and might lead to more time spent by humans tending the fields and orchards.

Non-humans. Non-humans include a wide variety of life forms: animals, fungi, and other living beings such as plants, trees and even whole ecosystems

27 Jungell-Michelsson and Heikkurinen 2022; Bocken and Short 2016.
28 Schumacher 1973.

seen in a holistic way (rivers, lakes, mountains).[29] These groups of beings are usually considered only partly or marginally by companies. Examples of existing considerations of non-humans include practices directed at animal rights and protection, e.g., where a company produces vegan products or donates to animal charities. While such considerations of non-humans are important, they are very limited in their scope. To broaden the scope of considerations towards non-humans, a business can pursue sufficient production as well as sufficient land use for other ethical reasons beyond the use of animal products. The core premises of sufficiency include acknowledging that human-made and natural capital are complementary (not substitutable), that human-induced metabolism needs to slow down, and that economic activity requires a degree of altruism.[30] In practice, a business can, for instance, actively seek ways to allow non-humans to thrive in their habitats rather than exploiting those habitats for the purpose of their conversion into human-made artifacts and built environments. Production of both goods and services inevitably requires the use of resources, thus destruction of, or interference with, habitats of non-human beings. The only way to deviate from this is to produce sufficiently rather than pursue ever more.

The self. It has been argued that such unusual, if not outright radical or even utopian, business practices can only stem from radical shifts in worldviews arising from nurturing moral agency.[31] Such a shift concerns self-transformation of humans involved in businesses as well as self-transformation of business as a social entity reflected in its operations, including the notion of sufficiency.[32] For instance, owners and top managers of a business can strive for sufficient income rather than ever increasing income or an ever more luxurious lifestyle. Not only managers but all employees of business organisations can reflect upon the double perspective they possess. That is to say that they are also consumers even when they represent the producing business. Reflective questions may be asked, such as "If I want to cut down consumption, what would it require from the producing side, and vice versa?" This is the domain where business practices and principles meet in a dialogue with the personal principles and practices of business owners, managers, and employees. A business itself, as a community of humans, can intentionally strive for sufficient growth, i.e., grow to a limit rather than pursue endless expansion across

29 Nesterova 2021c.

30 Jungell-Michelsson and Heikkurinen 2022.

31 Jungell-Michelsson and Heikkurinen 2022; Nesterova 2021b, 2021c; cf. Bhaskar, 2012.

32 Nesterova 2021a.

TABLE 6.1 Relationships between degrowth business and sufficiency

Degrowth business dimensions	Relationship to sufficiency
Nature at large	Shortening of supply chains, withdrawing from exploitative activities in various parts of the world
Location	Sufficiency of location: localisation of production, consumption of both goods and services
Humans	Encouraging sufficient consumption, sufficient size of business, sufficient number of products
Non-humans	Sufficient production and land use to leave sufficient spaces for non-human beings
The self	(Self)transformation towards sufficiency rather than ever more, developing worldviews which promote moral growth rather than growth in the material (personal gain and profit)

various indicators.[33] This entails intentionally foregoing potential earnings and gains in power to contribute to something more elusive but undoubtedly more important, i.e., harmonious co-existence between humans and nature as well as within humanity.[34]

What becomes evident is that incorporating sufficiency as a principle for a business of deep transformations goes far beyond the practice of (eco)efficiency, a practice which is most commonly referred to in relation to more

33 Ibid.

34 This is where one needs to be careful with expressing optimism regarding change. Is such change at all possible? The question of such foregoing of what has become valued and valuable in our societies (monetary gain, power) is essentially a question of human nature which remains obscure. Will humans impose self-limitation to sustain life on Earth, especially life which they will never meet due to spatial (distance), temporal (lifetime) or scale related (e.g., consider insects, fungi, and even bacteria) constraints? This is hard to know. In different literatures different aspects of being human have been highlighted. The philosophy of metaReality emphasises human capacity for love, creativity, and freedom (Bhaskar 2002). Humanism highlights human goodness as well as evil (Tuan 2008, 2013). While it is safe to say that there is no universally agreed upon understanding of what human nature is exactly, it is also possible to assume that humans can intentionally infuse their life with meaning even in the situation where the world itself does not readily provide meaning (Camus 2005), where the meaning is lost (Naess 1989) or where the meaning offered is that of a capitalist system (Lefebvre 2014), i.e., the meaning of life is reduced to having rather than being (Fromm 2013).

ecological production. Efficient production entails approaching resource use frugally, deriving the maximum possible output from a certain input of materials and energy. Such practice does not contradict sufficiency, and in fact it can readily complement it. However, focusing solely on efficiency invites a narrow focus on, and possibility of, techno-optimism that we briefly discussed above. Sufficiency, as opposed to efficiency and technology, as a guiding principle spans across various domains, from considering a business's location to considering individuals' worldviews. It is not restricted to the processes of production.

It is important to highlight that there are many terms within the field of degrowth which are hard to define, such as "small scale" and the boundaries of location. It is equally challenging to define quantitatively what sufficiency is. Sufficiency is highly subjective and depends on one's personal and situational circumstances. What is sufficient to one person is not necessarily sufficient to another one. Likewise, what is a sufficient size for one business may differ to what another perceives to be sufficient. Opposing highly prescriptive, authoritarian, top-down approaches, such as assigning carbon allowances to individuals and businesses, leaves us with few options to help guide the definition of sufficiency. One approach to sufficiency may be supplementing this concept and value with other related ones such as smallness and human scale,[35] thus seeking sufficiency in relation to these. Some limits may be natural, e.g., planetary boundaries or provided by the nature of locations themselves. While such limits may indeed define what could be naturally sufficient (e.g., the use of local natural resources), the difficulty lies in understanding how careful and sufficient use of resources should be implemented and who would enforce and control this use without resorting to a dictatorial regime.

3 Sufficiency of Location

In degrowth scholarship, geography remains insufficiently included, and geographical concepts such as space, place and location often remain unaccounted for.[36] From a geographical perspective, it is essential to remember that things (events, phenomena) take place. They exist and unfold somewhere, in a constellation of physical as well as social structures. Such constellations are unique to specific locations. What is possible to produce sustainably in one location is different to what is possible to produce sustainably in another

35 Bookchin 1982; Schumacher 1973.

36 Nesterova 2022b, 2022c.

location. Some practices of production are even location specific. For instance, reindeer herding is practised only in a few countries (including the Nordics). Blueberry wine is best produced where locally and wildly grown blueberries are available, such as in Sweden and Finland. Any theorising of degrowth practices should therefore be linked in both theory and practice with the nature of location. However, a location should not be seen as a constraint, but rather a source of multiple possibilities. It guides what could be produced, reveals which materials are available and directs the rhythms of production. For instance, wine is typically known as a product of warmer climates and viticulture, but it has historically been (and is still today) produced also in the Arctic region—in this case from local berries, such as blueberries, lingonberries, and raspberries.[37]

The challenge with location is that it is notoriously difficult to define where a location ends and something else, such as the wider nature or the cosmos, begins. A location can be a village, a region, a country. When one talks about localised production, does it mean, for instance, the metropolitan area of Helsinki, the North of Sweden or Europe? Perhaps sufficiency of locations depends on the nature of an industry and there can be no single universal definition of what localised production is. For example, while the need for blueberries can be satisfied by foraging locally, the need for modern medicine cannot be satisfied by highly localised production or local dwelling.

A principle that can guide sufficiency of location is striving for small scale production wherever possible.[38] Trainer (2020) notes that small scale production can facilitate efficient and quick management by communities and eliminate the need for many activities of the value chains. Another principle is a quest for resources that can be procured locally. Capitalist societies erase our connection with locations and distract humans from the uniqueness of locations.[39] The whole world becomes a resource pool and a single, global market. In winter, humans in Northern Europe have come to expect fresh berries, and out-of-season availability thereof has become normalised in the capitalist society. Such possibility is afforded only by importing these berries from South America and South Africa, efforts which come with high energy costs and the use of resources such as plastic. Sufficiency of location entails living in harmony with the rhythms of nature instead, using methods to preserve local foods and minimising imports or limiting them to a region, for instance, to Europe.

37 Nesterova 2022d.

38 Trainer 2022.

39 Harvey 1996.

Naturally, a path to sufficiency of locations is, just like a degrowth business itself, a journey. Some types of production are easier to localise while others cannot be localised at all or can be localised only to some degree. For instance, Trainer (2020) provides an example of localisation of egg production which results in substantially reduced ecological impact as well as increased community bonding. However, localisation of production of modern medicine[40] or services such as railways which are, by definition, out-reaching is more challenging, if at all possible or even desirable. One place to start is to identify those goods and services which can be effectively produced locally and see consumption of local goods as a value in itself rather than a sign of deprivation. Meaning and happiness can be sought not in the domain of consumption and exploring ever greater ranges of options, but in sufficiency and connections with humans, non-humans, and nature.[41]

To avoid pursuing extremes, such as limiting production to a local town and its surrounding area, one place to start is a region as a location. Paech (2017: 484) notes the following: 'In between the two extremes of local subsistence and global division of labour, regional economies can be developed as another complementary supply system. Complementary currencies could stabilise the regional economy and bind demand to the region. Entrepreneurial advantages due to specialisation could be used, but would be based on de-globalised and work intensive technologies.' Hence, regionally localised production can be supplemented with other regionally localised and supportive activities such as introduction of complementary currencies and seeking more appropriate rather than high technologies, as well as increasing humans' participation in local industries. Some of those industries may include various forms of production where consumers and producers come together, such as community supported agriculture.

4 The Dynamics of Production and Consumption

A business as well as its customers exist within a complex constellation of various social structures. Such structures include economic, political, and cultural ones. They are deeply interrelated and can be both limiting and empowering for deep transformations. A business and its customers are inseparable from

---

40 Production of modern medicine is extremely complex and involves multiple, often global, processes even before the production itself takes place. See for instance Azzaro-Pantel 2018.

41 Naess 1989, 2002, 2016.

these structures and cannot be understood apart from them. Both businesses and consumers are influenced by, as well as influence, social structures, creating certain dynamics of production and consumption. The dynamic of production and consumption deserves particular attention when considering sufficiency. Previous literature has only to a limited extent connected consumers and producers to each other.[42] We argue that for sufficiency to become a guiding principle in business as well as in consumption, consumption and production need to be considered in tandem.

4.1 *Economic Structures, Sufficient Production, and Sufficient Consumption*

Producers and consumers are conventionally seen as the basic microeconomic actors that interact through money-mediated exchange of goods and services. This exchange, alongside production and consumption, are the foundations of what is commonly seen as the economic domain. Sufficiency can hardly be achieved if only one of these actors, either a producer or a consumer, pursues it. Both groups of actors need to work towards sufficient production and consumption via establishing a dialogue. Consider the food sector, for instance. Food producers and retailers (such as supermarkets) have the possibility to choose which food products are produced, marketed, and become available on the supermarket shelves. Consumers have few options but to buy what is offered in this way. On the other hand, food producers closely follow the movements and changes in practices and preferences of consumers. If sufficiency (rather than hedonic, excessive, and luxurious consumption) becomes a widespread phenomenon, producers would have no option than to adjust. This adjustment is best facilitated by communication whereby consumers and consumer groups can work together on, for instance, making sure that seasonal and locally produced food is available. Within the economic domain, businesses can deviate from creating demand for products which are hedonic and use advertising to help consumers step on the path of sufficient and more localised consumption.

4.2 *Political Structures, Sufficient Production, and Sufficient Consumption*

Political structures influence the dynamic between production and consumption. Critical political economy explicates interactions between capitalism and the society's dynamics, including the possibility of transformations.[43] In the capitalist society, both producers and consumers are encouraged to produce and consume more rather than sufficiently. Stepping away from capitalist

42 Jungell-Michelsson and Heikkurinen 2022.

43 Buch-Hansen and Carstensen 2021.

tendencies gives an opportunity to pursue sufficiency in both production and consumption. The dynamic between producers and consumers in a society stepping away from capitalism onto a degrowth path should become more dialectical, whereby both producers and consumers work towards a shared goal of needs satisfaction.[44] Politically speaking, how such transformation will unfold and where it will come from is challenging to tell. This may depend on the type of capitalist system in place.[45] In countries with liberal forms of capitalism (the UK, the US), consumers and their grassroots movements may encourage businesses to supply more local and durable goods which would enhance their sufficiency pursuits and practices.[46] In countries with state capitalism (e.g., China, Russia), encouragement may come from the government. For instance, governments can support local businesses producing food via low or no interest green finance schemes and subsidies.[47] Currently, it is the businesses which experience or promise economic growth that are commonly supported rather than those which satisfy genuine human needs.

4.3 *Culture, Sufficient Production, and Sufficient Consumption*

What underlies the process of orientating both production and consumption as well as the dynamic between them towards sufficiency, in our view, is a cultural shift of mindset and values. As we identified in the beginning of this chapter, central to such cultural shift is growth in moral agency which entails nurturing human qualities such as empathy, compassion, solidarity, care, kindness, self-transcendence.[48] Ultimately, businesses consist of human beings, and consumers are human beings. Likewise, the political parties as well as labour unions and various societal movements mentioned in the previous sub-section consist of human beings. The changed worldviews need to be orientated towards sufficiency of one's location and respect towards nature at large. Orientation towards location does not mean looking strictly inward. Indeed, foregoing exploitation of humans, non-humans and nature does not mean foregoing 'the cultural and intellectual riches of the larger world'.[49] While focusing on location it is entirely possible to appreciate the world beyond one's

44 Nesterova 2020.

45 Buch-Hansen 2014.

46 Buch-Hansen et al. 2024; Buch-Hansen 2014.

47 While we do not dive deeper into policies that can facilitate sufficiency orientated businesses and consumption, policies have been discussed in, for instance, Buch-Hansen et al. 2024.

48 For a more in-depth discussion, see Buch Hansen and Nesterova 2024.

49 Tuan 1996, p. 13.

location. As Tuan[50] notes, such appreciation may be cultural, intellectual, but it does not have to presuppose viewing the world and nature as a pool of economic resources.

Change in culture towards sufficiency can be nurtured within (but should not be limited to) existing movements such as voluntary simplicity, slow living, minimalism, and the zero-waste movement which can lead deep transformation by example.[51] Such movements (though not without their own problems) generally promote lifestyles based on sufficient consumption as well as more harmonious co-existence between humans and nature. Existing consumers with philosophies targeting ecological impact minimisation are counterparts to existing sufficiency driven businesses. However, it is important that trends and philosophies of sufficiency do not merely exist in isolated niches and within defined movements and spaces but reach out to wider audiences, such as through the political pathways identified above.

One way to more effectively engage with existing sufficiency practices in business is to see sufficiency in non-binary terms, i.e., avoiding viewing business as fully sufficiency-driven or not sufficiency-driven. More often than not, sufficiency practices are intertwined with other practices in a business. Noticing such practices and their diversity can enhance engagement with sufficiency and learning how sufficiency is and can be implemented.

5 Illustrative Cases

To illustrate how sufficiency features in businesses which are on diverse paths of deep transformations, we turn our attention to several cases from Northern Sweden, Finland, and England. These cases are ones our authorship team researched during our academic careers. Many of the cases are situated in the food sector, which is interesting from a sufficiency (as well as degrowth and deep transformations) point of view as food represents a basic human need (we cannot stop consuming and producing food) and is contextual and local in many ways. However, having said that production of food is aimed at satisfying a basic need of humans for nourishment, nuances remain. For instance, one can identify types of food production which are more aligned with needs satisfaction, while others with pleasure. Amongst the cases below, one may find a market garden selling organic vegetable boxes, which can be seen as satisfying a genuine need. Such firms can be seen as sufficiency cases in terms of both

50 Tuan 1996.

51 Nesterova 2024b.

satisfying needs and implementing sufficiency elements into their business models. Firms focusing on producing wine and chocolate can be seen as satisfying desires (or wants). Even though their products may not be aimed at satisfying genuine needs, these firms likewise implement elements of sufficiency. In what follows, our aim is to highlight the processual and nuanced nature of these businesses rather than label them as, for instance, degrowth or not degrowth. We thus place emphasis on the variety of manifestations of sufficiency rather than on a business being sufficient in one sense or another. While we link each case below to Table 6.1, it is important to note that all companies in their journeys consider nature at large in one way or another and include ecological considerations in their businesses.

5.1 *Wine Producer in Northern Sweden*

The first case is a wine producer in the North of Sweden. The company was established in 2018. Its wine was inspired by a product called "sun wine" which has been traditionally home-made in Sweden. Sun wine was made by individuals using a natural fermentation process: blueberries were placed in bottles with sugar and allowed to ferment in the sunlight. Commercial production of blueberry wine, however, entails using professional equipment and expert knowledge, while the ingredients are frozen blueberries, water, sugar, and yeast. Sufficiency manifests itself first and foremost in the consideration of the company's location, in that local berries suffice and are indeed a source of inspiration and a key ingredient in a product. The company utilises locally grown, wild blueberries and lingonberries for its products and does not plan to expand into more traditional grape wine production. The company aims to facilitate consumption of locally produced wine instead of grape wine which cannot be produced in Northern Sweden. Due to this desire of the company to encourage local and naturally grown berry wine consumption (rather than consumption of imported grape wine), the company aims to expand. Thus "sufficient size of business" (see Table 6.1) is currently not on the company's agenda, while "location" certainly is.

5.2 *A Firm Specialising in Vertical Hydroponic Agriculture in Northern Sweden*

The second case is a firm specialising in vertical hydroponic agriculture based in Northern Sweden. The company was established in 2021. It uses high technology to control the delivery of water, nutrients and for controlling the moisture and light in the space where lettuce and herbs are produced indoors via the technique called vertical hydroponic agriculture. This technique allows to grow mainly lettuce and herbs without the use of soil. Sufficiency is manifested in this company via the owner-managers practising lifestyles of voluntary simplicity and mindfulness which spills over their approaches to business ("the self" in Table 6.1). Sufficiency is also manifested via the desire of the owner-managers to focus on the location, to be able to produce local food ("location" in Table 6.1). However, the company plans to expand not only its production of lettuce and herbs but also venture into other initiatives locally, such as sourcing other types of local food from local farmers and making it available to broader audiences via using logistics and networks of a large local food producer.

5.3 *Chocolate[52] Manufacturer in Northern Sweden*

The company was established in 2013. The founder works directly with cocoa farmers in South America to source cocoa beans as well as to preserve old cocoa varieties. The company also works with local producers in Sweden to source a variety of berry powders and other natural ingredients from the North to flavour and decorate the chocolate. Due to the cocoa being grown far away from Sweden, the company's production is a mixture of the local and the far-away. While most of the raw materials travel far to participate in the process of production, the local is evident in the company's work with local volunteers, experts (e.g., chemists), and designers. Where sufficiency is most prominent is perhaps the focus of the company on a single key ingredient: cocoa. Such focus allows the company to monitor closely the quality and the whole process of production from growing to distribution as well as to develop unique chocolate products (bars with extracts from plants which grow in the North, cacao nibs, cacao powder) and services inspired by chocolate, such as tasting sessions ("humans" in Table 6.1).

5.4 *A Market Garden in Southern Finland Selling Organic Vegetable Boxes*

This case is a small-scale, organic farm in an urban setting in Southern Finland. The farm has been managed in its currents form since 2013 and is currently run by a young couple. The couple grows vegetables, potatoes, and herbs on open land and sells these directly to consumers in weekly vegetable box schemes. They serve around 50–100 local consumers who come and pick up their vegetables at the farm, as well as some nearby restaurant customers.

52 Chocolate is not a need, and cocoa is certainly not a local product in Sweden. While acknowledging this, we also want to emphasise the need to implement sufficiency within existing businesses as soon as possible, and understand how sufficiency is implemented currently, regardless of product.

This is done between approximately June to October, or as long as there are vegetables available. Sufficiency is manifested in several ways in these small-scale operations. The couple only sells what they have according to season and availability, the production is only as extensive as the area and nature of the farm allows, and the sale of products is limited to local customers that live close to the farm. Additionally, the managing couple has actively chosen to live a simpler life (they shifted from well-paid employment in larger institutions to farming) and earn only what is necessary to manage their family (which includes two children), but not more. In relation to Table 6.1, this case relates to "location" as well as "the self".

5.5 *A Company in Southern Finland Selling Local Game Meat*[53]

This company was established in 2013 to provide consumers as directly as possible with locally hunted game meat. The founder's mission was to increase the appreciation for Finnish game meat and local Finnish food culture in general ("location" in Table 6.1). He built a small slaughterhouse at the farm where he grew up and lives at the countryside in Southern Finland and uses this as a hub to which local hunting teams can bring their catch (mostly moose and deer). The company then takes care of the meat and cuts up or processes it into meat parts and products that can be sold to consumers. Selling is done directly to consumers through the company's own webpage as well as through some selected stores in the metropolitan area. While the consumption of meat is questionable from a general sustainability as well as sufficiency point of view, the founder emphasises that game can be perceived as a rather sustainable meat alternative. Game meat has good nutritional values, and ethical hunting (defined by the company as selective hunting and gamekeeping) can support a thriving and balanced habitat. Additionally, the availability of game is restricted to hunting seasons and regulations, making it a product for "less, but

53 For our authorship team, inclusion of this case was in itself a journey and a subject to discussion. We are aware of the unfolding discourses on the relationship between climate degradation and animal products consumption. Ecological degradation means rethinking our consumption of animal products. It also encourages deeper considerations of animal rights and the nature and practice of animal husbandry (see e.g., Kortetmäki and Oksanen 2021). However, living in the Nordics makes us aware of the complexity and nuance in relation to meat consumption. Meat consumption (e.g., reindeer meat) has cultural and subsistence related roots, for instance, in the Arctic region where vegetables and fruits do not grow and cannot be grown in abundance due to the harsh and cold climate. Historically, the reliance on meat and fish products has been prominent. We invite our readers to consider this.

better" consumption. All of these can be considered manifestations of sufficiency in business.

5.6 *Natural Dyes Producers in England*

The final case we will refer to in this chapter is a small natural dyes producer based in England. The company was established in 2007, and apart from its main strand of operation (natural dyes), it also sells natural fibres and natural hand-made paper. The company is operated by a family of two (husband and wife) and is occasionally helped by a person who is a sole trader. Sufficiency manifests in this company profoundly and on many levels (primarily in relation to "humans" and "the self" in Table 6.1). The family running the company states that they do not want much in terms of their personal consumption and material wealth. Neither do they wish to expand their business and hire more people. They see growing demand for their products as a hindrance to their way of living and their work-life balance rather than something to celebrate. They would prefer other businesses to engage in similar activities (such as natural dyes production) rather than grow their own business. Free time, education (e.g., giving lectures, teaching others about natural dyes vs conventional dyes), art and craft are very important to them, and are not the aspects of life they would like to give up on to run a larger and steadily growing business, something that would conventionally be seen as success.

6 Conclusion

Once Murray Bookchin (1982: 134) noted that '[w]hat makes capitalism so unique is the sweeping power it gives to economics: the supremacy it imparts to *homo economicus*', a greedy, selfish, and materialistic "human". Within this strand of thinking, humans are happy when their consumption of goods and services increases. Moreover, humans are supposed to be indifferent to the fact that their increasing consumption or affluence, combined with ever more complex and resource intensive technology and growing population (IPAT) leads to the destruction of life on Earth, their home. Neither mainstream economics nor capitalism can effectively handle human nature and human pursuits. Above a certain threshold, increasing consumption is a burden, not a source of happiness. Leopold[54] reminds us that if an increase in something contributes to higher wellbeing, it does not mean that indefinite increase

54 Leopold 1949.

likewise contributes to wellbeing. Thinking in terms of sufficiency of goods and services is fruitful for both consumers and producers. After all, both groups are comprised of human beings with capacities transcending the pursuit of material goods, money, and status. It is impossible to tell what comes first or what should come first, sufficient production or consumption. Perhaps it is not the right question to ask, and it is working together towards sufficiency in both production and consumption that counts. We suggest that redirecting both production and consumption towards sufficiency is necessary. These are inter-related processes – if not one process – of a deep transformation in our relationships with the world. In this chapter we focused on sufficiency as a potential principle for business. We propose that degrowth business, i.e., a business of deep transformations, is a business of sufficiency. Apart from discussing how sufficiency may manifest itself in business operations, we highlight the need to contemplate sufficiency of locations, to localise production as much as possible. Yet, neither business of sufficiency nor localised production are stationary. Both should be seen as processes.

References

Azzaro-Pantel, C. (2018) "Chapter 1 – New Product Development and Supply Chains in the Pharmaceutical Industry," *Computer Aided Chemical Engineering* 41, 1–26.

Bhaskar, R. (2002) *meta-Reality: The philosophy of meta-Reality. Volume I: Creativity, love and freedom*. London: Sage.

Bhaskar, R. (2012) *The philosophy of MetaReality*. London: Routledge.

Bocken, N.M.P./Short, S.W. (2016) "Towards a sufficiency-driven business model: experiences and opportunities," *Environmental Innovation and Societal Transitions* 18, 41–61.

Bonnedahl, K.J./Heikkurinen, P. (2019) "The case for strong sustainability," in Bonnedahl/Heikkurinen (2019) *Strongly Sustainable Societies: Organising Human Activities on a Hot and Full Earth*. London: Routledge, 1–20.

Bookchin, M. (1982) *The ecology of freedom: The emergence and dissolution of hierarchy*. Palo Alto: Cheshire Books.

Buch-Hansen, H. (2014) "Capitalist diversity and de-growth trajectories to steady-state economies," *Ecological Economics* 106, 167–173.

Buch-Hansen, H. (2021) "Modvækst som paradigme, politisk projekt og bevægelse," *Nyt Fokus Fra Okon. Vækst Til Bæredygtig Udvikl.* 17, 6–9. Available online: http://www.nytfokus.nu/nummer-17/modvaekst-som-paradigme-politisk-projekt-og-bevaegelse/ (accessed 5 January 2022)

Buch-Hansen, H./Carstensen, M. B. (2021) "Paradigms and the political economy of ecopolitical projects: Green growth and degrowth compared," *Competition & Change* 25(3–4), 308–327.

Buch-Hansen, H./Nesterova, I. (2021) "Towards a science of deep transformations: Initiating a dialogue between degrowth and critical realism," *Ecological Economics* 190, 107188.

Buch-Hansen, H./Nesterova, I. (2023) "Less and more: Conceptualising degrowth transformations," *Ecological Economics*, 205, 107731.

Buch-Hansen, H./Nesterova, I. (2024) "Deepgrowth: Self-transformation towards harmonious being," *Ecological Economics*, 224, 108281.

Buch-Hansen, H./Koch, M./Nesterova, I. (2024) *Deep transformations: a theory of degrowth*. Manchester: Manchester University Press.

Camus, A. (2005) *The myth of Sisyphus*. London: Penguin Books.

Drews, S./Antal, M. (2016) "Degrowth: A "missile word" that backfires?" *Ecological Economics* 126, 182–187.

Fromm, E. (2013) *To have or to be?* London: Bloomsbury.

Hankammer, S./Kleer, R./Mühl, L./Euler, J. (2021) "Principles for organizations striving for sustainable degrowth: framework development and application to four B Corps," *Journal of Cleaner Production* 300, 126818.

Harvey, D. (1996) *Justice, nature and the geography of difference*. Oxford: Blackwell.

Heikkurinen, P. (2018) "Degrowth by means of technology? A treatise for an ethos of releasement," *Journal of Cleaner Production* 197, 1654–1665.

Heikkurinen, P./Ruuska, T. eds. (2021) *Sustainability beyond technology: Philosophy, critique, and implications for human organization*. Oxford: Oxford University Press.

Jungell-Michelsson, J./Heikkurinen, P. (2022) "Sufficiency: A systematic literature review," *Ecological Economics* 195, 107380.

Kortetmäki, T./Oksanen, M. (2021) "Is there a convincing case for climate veganism? Is there a convincing case for climate veganism?" *Agriculture and Human Values* 38, 729–740.

Lefebvre, H. (2014) *Critique of everyday life*. The one-volume edition. London: Verso.

Leopold, A. (1949) *A Sand County Almanac: And Sketches Here and There*. Oxford: Oxford University Press.

Naess, A. (1989) *Ecology, community and lifestyle*. Cambridge: Cambridge University Press.

Naess, A. (2002) *Life's philosophy: Reason and feeling in a deeper world*. Athens: The University of Georgia Press.

Naess, A. (2016) *The ecology of wisdom*. London: Penguin Classics.

Nesterova, I. (2020) "Degrowth business framework: Implications for sustainable development," *Journal of Cleaner Production* 262, 121382.

Nesterova, I. (2021a) "Small firms as agents of sustainable change," *Futures* 127, 102705.

Nesterova, I. (2021b) "Small, local, and low-tech firms as agents of sustainable change," in Heikkurinen/Ruuska (2021) *Sustainability Beyond Technology: Philosophy, Critique, and Implications for Human Organization.* Oxford: Oxford University Press, 230–253.

Nesterova, I. (2021c) "Addressing the obscurity of change in values in degrowth business," *Journal of Cleaner Production* 315, 128152.

Nesterova, I. (2022a) "Degrowth perspective for sustainability in built environments," *Encyclopedia* 2, 466–472.

Nesterova, I. (2022b) "Business of deep transformations: How can geography contribute to the understanding of degrowth business?" Geography and Sustainability 3(2), 105–113.

Nesterova, I. (2022c) "Being in the world locally: Degrowth business, critical realism, and humanistic geography," Frontiers in Sustainability 3, 829848.

Nesterova, I. (2022d) Arctic winescapes. Available at: https://criticalarcticstudies.com/2022/04/11/arctic-winescapes/ (accessed 01 March 2023)

Nesterova, I. (2024a) Degrowth, depth and hope in sustainable business: Reflections from Denmark, Finland and Sweden. London: Routledge.

Nesterova, I. (2024b) "Being of deep transformations: A personal journey inspired by Clive L. Spash." Environmental Values 33(2), 122–138.

Paech, N. (2012) Befreiung vom Überfluss: Auf dem Weg in die Postwachstumsökonomie. München: Oekom.

Paech, N. (2017) "Post-growth economics," In Spash (2017) *Routledge Handbook of Ecological Economics: Nature and Society.* London: Routledge, 477–486.

Sandberg, M. (2021) "Sufficiency transitions: A review of consumption changes for environmental sustainability," *Journal of Cleaner Production* 293, 126097.

Schumacher, E.F. (1973) Small is beautiful: A study of economics as if people mattered. London: Blond & Briggs.

Trainer, T. (2022) "A technical critique of the Green New Deal," Ecological Economics 195, 107378.

Tuan, Y-F. (1996) *Cosmos and hearth: A cosmopolite's viewpoint.* Minneapolis: University of Minnesota Press.

Tuan, Y.-F. (2008) *Human goodness.* Madison: The University of Wisconsin Press.

Tuan, Y-F. (2013) *Landscapes of fear.* Minneapolis: University of Minnesota Press.

Whiteman, G./Cooper, W. H. (2011) "Ecological sensemaking," *Academy of Management Journal* 54 (5), 889–911.

PART 3

Sufficiency in Technology

CHAPTER 7

Getting Our Bearings after Oil: Technology, Sufficiency and Types of Knowledge

Tere Vadén

1 Introduction: PAT in the Black Light of Oil

During the past 200 years or so, the belief that humanity is independent of nature or even master over it has become more prevalent. The geological effects of the supposed mastery have been led to the proposal of a new era, Anthropocene, and the intensified pace of action has been characterized as the Great Acceleration.[1] However, the belief contains a supreme irony: the supposed independence from nature relies on a one-time, exhaustible natural endowment, namely fossil fuels. The technological prowess and the material affluence of modernity rely on the work done by fossil fuels. This work is an order of magnitude greater than the maximum muscle labor that currently living humans (and working animals) could theoretically muster.[2] The supposed distance or independence from nature is produced by using particular types of natural resources—fossil coal, oil, and gas.

The irony is "black" in the sense that by their very nature fossil fuels, especially oil,[3] function in ways that hide their essential role. Oil extraction has presented unique net energy values, oil has huge energy density, it can be easily transported and stored. These characteristics have made possible the separation of industrial production from any particular time (daylight, the seasons) and place (sources of water power, centers of consumption), so that capital has

---

1 Crutzen, P. & Stoermer, E. (2000) The 'Anthropocene.' *IGBP Newsletter* 41; Steffen, W., et al. (2015) "The trajectory of the Anthropocene: The Great Acceleration." *The Anthropocene Review*, 2 (1).

2 Smil, 2008.

3 While fossil fuels (coal, natural gas, oil) as a group share many essential characteristics, such as geological origin and high energy return on energy investment, they also differ materially, so that a social metabolism relying on coal is in some crucial respects different from one relying on oil, see Mitchell 2011. In the following, I will be mostly referring to the socio-metabolic characteristics of oil, as its use did and still does more physical work than coal and natural gas, and as it is the most voluminous resource for petrochemistry and use in internal combustion engines.

 | DOI:10.1163/9789004746404_008

sought cheap human labor and lax legislation where it can be found.[4] Modern consumption is crucially dependent on production that (mostly) happens far away. The mode in which oil connects the world, is the mode of making and keeping distant.[5] A similar distance characterizes the experience of technology; its energetic conditions are obscured even in the I = PAT formula itself.

A second-order "blackness" is even more striking. Philosophers and critics of the Zeitgeist have analyzed minutely many features of modernity—ownership of the means of production, division of labor, the role of technology, mass society and so on—without noticing that these are also dependent on the one-time endowment of fossil fuels. True, some astute observers, such as Elmar Altvater, Andreas Malm, and Amitav Ghosh, have pointed out how well the properties of fossil fuels, and especially oil, fit those of particular power relations. Altvater (2007) coined the term "fossil capitalism" to describe the way in which fossil fuels facilitate capitalist economies. Andreas Malm (2016) has famously argued that the ascendancy of coal powered steam machines over water power in the 18th and 19th century Britain was not due to price, as often assumed, but due to the fact that by using steam power, the capitalists were able to appropriate a bigger part of labor than by using water power. Amitav Ghosh (2021), in turn, has noted how fossil fuels and colonialist power fit in "uncanny" ways.

A wide set of conditions—social, economic, spiritual, technological—must be favorable for anything like modern societies of overconsumption to flourish. Here I want to concentrate on just one aspect, an epistemological one. Fossil capitalism pushes aside, obstructs, and usurps economic relations that are not of the capitalist kind, while arguably also relying on them. Analogously, it can be expected that fossil capitalism has its favored systems and types of knowledge, and pushes aside, obstructs, and usurps different kinds of knowledge and knowledge systems. If this is the case, we can expect to find symptomatic epistemological effects, specifically distortions and blind-spots concerning energy, in the knowledge about fossil capitalism itself, in its "self-understanding," so to speak.

4 As Marx points out in Chapter 33 of *Capital*, with the example of Mr. Peel who moves from Britain to Australia with necessary workforce and productive capital in tow but still fails to set up capitalist production, capitalism needs certain social relations (including, typically, a police force and the impossibility of subsistence livelihoods), not just the workers and the capital.

5 For more on this phenomenon of connecting-by-keeping-distant, see Salminen and Vadén (2015).

Let us take an example. The famous I = PAT formula maintains that the Impact to ecosystems by a human population are a product of the size of the Population, its level of Affluence and Technology. All of the terms on the right side of the equation, Population, Affluence, Technology, are directly affected by the material and energetic base of the society. As economist Steve Keen has pithily observed, "labor without energy is a corpse, and capital without energy is a sculpture".[6] In fossil capitalism, PAT is elementally dependent on fossil fuels. However, the role of fossil fuels is, at best, implied as in the equation, and, at worst, wholly ignored. The wager in the following is that a part of the uncanniness noted by Altvater, Ghosh and others derives from epistemological effects of fossil fuels, the blacking-out it has performed on knowledge. "A fog of war" emanating from fossil fuels permeates what we (seem to) know of I = PAT, and, conversely, any sustainability and sufficiency beyond catastrophic levels of Impact.

The recognition of this "blackness" necessitates a reappraisal of what modern technology and affluence are, and what are epistemological conditions of sufficiency. Most obviously, much of the credit (or blame, as the case may be) given to technology actually belongs to fossil fuels, and most of all, oil.[7] Without energy technology does not work. There is an urgent double-task at hand: first, to reappraise the modern experience of technology and affluence in the black light of oil and, second, to take that reappraisal seriously with regard to the approaching end of the fossil fuel era.

2 Describing the Failure

It would be fair to say that in popular and professional discourse alike, the idea is that current science and science-based understanding of the world is the best ever, quantitatively and qualitatively, likewise for the technological manipulation and control of the human environment. Modernity might even be defined as the period, where each century, decade and year sees that its view of the world is somewhat better than the view of any preceding century, decade, or year. Yes, backsliding might happen here or there, yes, mistakes are made and need to be corrected, but on the whole, it is thought that at any

6 Keen, S., Ayres, R., and Standish, R. (2019) "A Note on the Role of Energy in Production". *Ecological Economics*, Volume 157.

7 For a longer discussion, see Vadén, T. (2021) "What Does Fossil Energy tell us About Technology?" in P. Heikkurinen and T. Ruuska (eds.), *Sustainability Beyond Technology*. Oxford University Press.

given moment the prevalent science-influenced view of the world is the most unprejudiced and most correct it has ever been, and the technology better than ever.

This "progressivist" description of science does not go only for scientific knowledge—where "science" is understood as "scientific observations, facts and theories"—but also for science as a pursuit—scientific methods, institutions, and the philosophy of science embedded in scientific practices. That is, the progressivist idea is that not only are current scientific facts the most precise ever, but also the way in which we do science and the way in which science informs society is the best ever. Again, there is wide variation the world over in scientific institutions and the role and authority of science in society, but on the whole, the belief is that that the current era is the most science-imbued ever.

A fortiori, the progressivist description goes for the science called economics, as well, even though here more enlightened discussions around the political nature of science can sometimes be found. For decades, the theoretical and practical promise from the science of economics and the public use of economics as ground of argumentation in politics has been that a desirable and achievable goal for societies is continued economic growth.[8] Even today, it is hard to find in OECD countries political parties in parliament that would not rely on the idea of continued economic growth in their policies, programs, and rhetoric.

However, at the same time we know—with as much scientific consensus as we know anything—that fossil capitalism will end within a couple of decades. As the reports of the IPCC (e.g., 2018) with increasing emphasis explain, either it will end as we voluntarily move to other modes of energy production (modes and transformations which, as we also scientifically know, will mean deep socio-economic changes) or emissions will force a climate crisis that forces rapid involuntary simplification of organized societies. One way or the other, or most likely in a mixture of ways around the globe, the era of continuous economic growth in terms of fossil capitalism will end.

This means that to the best of our scientific knowledge, the era of fossil capitalism will last roughly 200 years. It is important to stop for a moment to appreciate this fact. In terms of continuity management, a run of 200 years is a massive failure. There are literally thousands of civilizations, cultures or societies that have stayed their course for much longer. There are even several civilizations or cultures that have managed to persist for thousands of years

8 Schmelzer (2016) provides a helpful summary.

without destroying their ecological basis. In a typical overview of human history, a cultural formation lasting 200 years does not even get a mention, unless its somehow disrupts the narrative of a more prominent cultural legacy.

We also know what the disruption is in this case. Fossil capitalism has already destroyed species, ecosystems, livelihoods, and cultures, and will still destroy much more. Ecological changes such as global warming, extinctions and plastic pollution will persist in the natural and historical records for millennia; from a human perspective forever. These effects and their socio-economic consequences will persist quite a lot longer than the phenomenon of continuous economic growth, as defined in fossil capitalism.

And yet: these disruptive effects are an accident. They are not the intentional products of fossil capitalism, but something that presumably "just happened" on the way, as collateral damage of things that fossil capitalist economy was intentionally doing and aiming at. So, yet another uncanny observation: the most prominent and lasting consequences of an era that sees itself as having the best understanding and scientific knowledge of the world, have happened by accident.[9]

The irony is sickening: A culture where one of the central pillars is a promise of continuous economic growth turns out to be almost entirely without skills of continuity management. An era that sees itself as the pinnacle of dispassionate knowledge production and verified understanding, produces immeasurable destruction by accident. How did a large portion of humanity become so unskilled? How do we account for the gap between the self-understanding of fossil capitalism and the accompanying era of modernity (goal: continuous economic growth, tool: best scientific understanding and technology applied vigorously), and its results (a run of ca. 200 years, megadeath by accident)?

2.1 *'A car doesn't care where the gasoline comes from'*

Let me start with an anecdote that happened long before the current Russian attack on Ukraine in 2022. Somewhere on social media, I wrote that when a Finnish motorist fills the tank, the gasoline most likely comes from oil fields in Russia, most likely from the areas inhabited by Finno-Ugric peoples (linguistically related to Finns), and oppressed by extractive industries and pro-Russian policies. My point was that by the simple act of driving and

9 There are obviously many levels on which the disconnect between the best scientific knowledge and some of the worst results in terms of continuity management and ecology can be analyzed. Certainly, on the political level, the prominence of (a certain school or paradigm of) economics is one of the reasons. Whatever the reason for the disconnect, it does exist, and as such, does also have epistemological effects.

filling up the tank, a Finnish motorist was supporting questionable activities, in Russia and by Russia. A reply pointed out a fact in a tone that nicely fits with the tenor of so-called new materialism: 'the car does not care where the oil comes from'. And that is the case: in refineries different oils with different qualities from different sources are, indeed, processed so that the end product, gasoline, is standardized for use in internal combustion engines all over the world. Car culture could not exist without this standardization. The driver does not have to know the providence of the gasoline. In fact, in many cases the motorist cannot know the providence, at least not without extensive time- and labor-consuming research.

True, there are "general" reasons, reasons not having to do with oil for the standardization. Many products on the market are standardized so that the origin of the raw materials used in them is obscure. Benefits of mass production, division of labor, consumer cultures, advertising and many other factors drive this kind of standardization. However, there are also "specific," oil-related, reasons to the abstraction.

The first is the obvious one: many of the features behind "general" standardization are actually dependent on the massive energy surplus that fossil fuels and specifically oil have provided for fossil capitalism. Without fossil fuels mass production and global division of labor would not be what they are now. In this sense, the "general" drive towards standardization is actually, in part, a "specific" effect of fossil fuels.

The second reason is the one aligned with so-called new materialist theories. The sheer amount of chemically identical hydrocarbon molecules (to be distilled into different grades such as diesel fuel, gasoline and kerosene) has made possible a global regime of production and consumption that simply would not be possible on the basis of burning other materials, such as wood or other biomass.[10]

This points to one crucial locus of the deskilling. The experience and skill of "filling up a tank" does not require knowledge of the origin of the gasoline, and is remarkably similar around the globe. Contrast this to other modes of transport. Imagine you are the captain of a sailing ship. Can you say that "the ship

10 Again, here oil is different from coal and natural gas, both of which can also function as resources for different petrochemical products, but typically with much less efficiency (energy-wise) than oil. The question of what kind of levels of production and consumption are possible on the basis of so-called renewable modes of energy production is a contested one. However, it is generally agreed that the net energy (or "energy return on energy investment," EROEI), for renewable sources is lower than that for fossil fuels, meaning that a larger portion of all work will have to go into energy production, diminishing energy available for other purposes. See, e.g., Rye & Jackson (2018).

does not care where the wind comes from" or something to a similar effect? Probably not. In fact, knowledge of the winds and weather systems and their connections to different localities, seasons and observable signs would, most likely, be a point of professional and individual pride. The same goes, if possibly with a lesser degree, for a steam ship. A captain of a steam ship would be particularly interested in the qualities and origin of the wood or coal that the boat is using, as the qualities in part determine the performance under steam and the need for maintenance.[11]

The lesson from the anecdote is that both because of "general" and "specific" standardization, oil deskills and fails to provide a material anchor for context-specific knowledge to accumulate. Thus, filling up the tank of a car is a paradigmatic case of the black sun of oil, de-lighting fossil modernity.

2.2 *Nafthism*

As noted above, the deskilling and unmooring of knowledge produced by oil does not concern only everyday experience, but also second-order pursuits, such as Zeitgeist analysis and critical philosophy. To take a few examples, when Marx and Engels predict in the *Communist Manifesto* that "all that is solid melts into air" they correctly describe a feature of modern experience, but fail to notice that this melting is made possible by fossil fuels. A capitalism running on water power might have been possible, and maybe in some senses worse than fossil capitalism, but the melting would quite literally have been slower. Likewise, when Heidegger (1954) says that contemporary technological or calculative understanding of Being encounters everything as raw material to be used, he fails to notice that material is raw only in the eyes of work, and that currently most of the work is done by fossil fuels. Without energy, without work, technology and matter can sit side by side, and nothing happens. Like Marx and Engels, Heidegger fails to note a necessary condition of his—as such—perceptive description. The failure is somewhat more damning for Marx and Marxists, who often pride themselves over keeping an eye on material realities.

During the past few decades, it has slowly dawned on the proponents of continuous economic growth that the use of fossil fuels on an industrial scale is producing environmental damage that threatens the fruits of industry. The obvious reply has been the notion of "green growth," predicated on the concept of decoupling, as stated by the OECD: "Decoupling environmental pressures from economic growth, with a view to ensuring that continued economic

11 The particularly energy rich coal from Wales has been cited as one source of British naval superiority in the run-up to the First World war, see, e.g., Gray (2017).

growth is accompanied by enhanced environmental quality, is needed."[12] The idea is, conceptually, impeccable. Unfortunately, practice has been more difficult. There is no empirical evidence of sustained, quick, and wide enough decoupling in the global economy.[13] The hope of decoupling relies on the idea of economic substitutivity. If there is shortage of a given resource in an economy, its price will rise thus giving incentive either to produce more of the resource, or eventually to find a substitute for the resource. The mechanism of supply and demand will via prices signal the need, and a substitute will be found.

Again, while the abstract description has obvious merit, the concept of substitution also misses something crucial. Particularly, energy is not just a resource on the market, but a precondition for the existence of a market in the first place. This is true on a general level: any market transaction takes energy. More specifically, an economy dependent (let us say for ca. 80 percent of its primary energy need) on fossil fuels is dependent on the specific properties of fossil fuels (such as transportability, storability, flammability, etc.), not just on the abstract amount of kilowatt hours or joules contained in those fossil fuels.

To sum up: In many cases and on many levels, we find notions that implicitly or explicitly miss the crucial importance of fossil fuels. Let us call this phenomenon *nafthism*: A description or concept of A is nafthist, if that description portrays A as independent of nature while, in fact, the very (illusion of) independence is dependent on the existence and use of masses of high-quality hydrocarbons in nature. Nafthist descriptions and notions are prime examples of the way in which the black sun of fossil fuels works, producing epistemological effects by masking the crucial role of fossil energy. Nafthism seems to be historically specific in somewhat similar ways as what Heidegger calls the technological understanding of Being: it permeates ways of experiencing and thinking as something "normal," thereby characterizing a whole era.

The nafthist oblivions correspond strictly to nafthist surprises, when the epistemologically masked appears via a detour. The best known of the surprises is climate change. It is worthwhile to do precise climate modeling and to appreciate the different effects of different greenhouse gases. Without global awareness, it would not be evident that gigatons of matter are being burned. And without global climate science, the effects on climate could not

12 OECD (2001) OECD Environmental Strategy for the First decade of the 21st Century. http://www.oecd.org/environment/indicators-modelling-outlooks/1863539.pdf, p. 11.

13 Vadén, T., et al. (2020) "Decoupling for ecological sustainability: A categorisation and review of research literature," *Environmental Science & Policy*, 112.

me monitored and understood. However, at the same time, raw common sense is enough to understand that if gigatons of matter are burned within decades, some consequences will ensue. Even in scientific terms, the basic mechanism of the greenhouse effect was described already in the late 19th century. Still, after 30 years of global climate conferences, the reality and urgency of climate change are having a hard time penetrating fossil capitalist consciousness.

The proliferation of plastic pollution, in ever smaller granularity, is maybe even better example of a nafthist surprise. If the industry produces gigatons of non-biodegradable materials that are not systematically collected and taken care of, then obviously that matter is going to end up in material flows everywhere, from oceanic gyres to human blood flows. There really can be no genuine "surprise" in the matter. Whatever degree of unexpectedness is connected to ubiquitous plastic pollution, it is a product of nafthism.

3 Wager: Lack of a Specific Kind of Knowledge

There are, to be sure, good grounds for the fossil capitalist world to believe that it has in its possession the best and most comprehensive set of scientific knowledge. Simple statistics are enough to tell that there have never been more scientists, more scientific institutions, more scientific articles and so on. This is an important clue. The problem producing the gap between the self-understanding and the subpar performance in continuity management is not due to the lack of doing science or of generating scientific results. In terms of the number of facts and explained phenomena, contemporary fossil capitalist societies really do possess the most abundant trove of scientific knowledge and the most advanced technology. Rather, the problem is in the role of knowledge, or the structure of the knowledge system as a whole, or the relations between knowledge production and the use of knowledge in the wider society—all of which I will for the present purposes see as problems of the "kind of knowledge", of the nature of knowledge.

From the other side of the gap, we know that some human civilizations, communities, or cultures have been able to maintain their dynamic identity for hundreds and even thousands of years without destroying the ecological basis they rely on. While there is not one example of a modern (fossil) way of life that would have been ecologically sustainable, there are multiple examples of human cultures that persisted for centuries or even millennia sustainably, and might, in some cases, have disappeared not because of self-induced ecological disaster but for other reasons, such as outside aggression, natural catastrophes and so on.

This gives a crucial clue. Let us imagine that there was a people living sustainably in some remote area, practicing a subsistence economy dependent on local natural resources. Then, accidentally, all of the individuals of the nation suffer a bout of acute amnesia relating to everyday practices. They do not anymore remember what they did for a living, how they gathered food, how water was managed, how shelter provided, and so on. Faced with an impasse, they turn to their knowledgeable and impressively wealthy fossil capitalist neighbors and ask how does one live sustainably in the natural area that they are occupying. The neighbors would certainly be able to provide a lot of peer-reviewed information on animal and plant stocks, weather patterns, historical customs of gathering, hunting, and fishing, and so on, together with examples of leap-frogging technology. But would they be able to point the way, show by living example that this is how it is done? Most likely not. It is quite telling that there exist no example of a society or culture that has lived or lives ecologically sustainably (i.e., in a way that current environmental science would call ecologically sustainable) and does that by using, applying or "having" the concept of "ecological sustainability".[14]

Clearly, the gap between the self-understanding and the actual performance of fossil capitalism is not produced by a quantitative lack knowledge, but through a lack of particular type of knowledge, more precisely of the kind of knowledge that would support continuity management and non-destruction of nature. Following this clue, we can trace some outlines of the missing type of knowledge by looking at what characterizes cultures that have been better at continuity management, and have succeeded in not destroying their ecological basis in mere 200 years.

Ecological non-destruction implies skillful coping[15] in a bioregion (or several bioregions), where the animal and plant species, features of landscape and patterns of weather are well known, so that good years and bad years, migrations, extreme events and so on are readily recognized and reacted to. In such a

14 The only possible examples may exist in contemporary Cuba; see Saunders & Luukkanen (2021). The point is not to downplay the experience and skills of Cubans, or of the other hundreds of millions of people living within the boundaries of one Earth and with good social sustainability in countries like Costa Rica, Vietnam, and so on, see, e.g., Hickel (2020). It is certainly possible to create living examples of socially and ecologically sustainable lives that are inspired both by scientific knowledge of ecology, economics and so on, and by non-Western and traditional ideas of social relations and nature. The point is rather that these people would likely describe their ways of life as *alternative* with regard to the model of fossil capitalism.

15 For the concept of "skillful coping," see Dreyfus (2014).

situation, continuity management can be embedded in many or most aspects of life, fractally, on many levels. Continuity management may happen explicitly in terms of rules of usufruct and limitations, or implicitly in practices like taboos and marriage systems, and everything in between. Studies of traditional ecological knowledge (TEK) contain examples of continuity management embedded in everything from customs of child-rearing to architecture and infrastructure.[16] Typically, the relationship of humans to non-human elements in the environment is characterized by respect, gratitude, and kinship, even while antagonism is also present.

Could this kind of continuity management be conscious, i.e., could it be an explicit goal in how those societies or cultures were or are run? In the absence of evidence to the contrary, we may suppose so. In their book *The Dawn of Everything,* David Graeber and David Wengrow (2022) put forward a detailed argument according to which the encounter of indigenous North American critique of Western inequality and unfreedom was an important impetus for so-called Enlightenment thought in the West. Graeber and Wengrow face the expected counterargument that such a sequence is unlikely as the natives of North America did not have the institutional structures for sustaining political and philosophical thought of the Enlightenment kind by pointing out that there is every reason to think that the indigenous population was intellectual equals with their European contemporaries. As they point out, it would be quite simply racist to suppose otherwise.

The same argument applies here. There is no reason to suppose that continuity management would not have been (or is not) a goal of how ecologically sustainable non-Western societies were (or are) run. To suppose otherwise would be simply racist. The argument stands even if these societies do not explicitly refer to concepts that would systematically map to scientific concepts of ecological sustainability and the like. The fact that such concepts and mappings may be missing does not show that goals like those of ecological sustainability and cultural continuity are missing. Rather, it points to the observation that the type of knowledge in which these goals are embedded are of a different type.

Often the elements of TEK (and knowledge regimens in contemporary ecologically and socially sustainable societies) are not divided into the categories recognized as domains of different scientific disciplines or even as domains of different natural kinds (i.e., groupings or categorizations that supposedly reflect the actual world, not just human preoccupations) in the scientific world

16 Gibson, T. and Sillander, K., eds. (2011) *Anarchic Solidarity. Autonomy, Equality, and Fellowship in Southeast Asia.* Yale University Press.

view associated with fossil capitalism. For instance, material, social and spiritual features are combined in undividable elements.[17] Crucially, these elements form a whole where the meaning of the cultural life, and the meaning of the life of the individuals, is tied together with its ecological sustainability. The point is not to first establish boundaries of sufficiency, and then to enjoy life within those boundaries in the way in which each individual sees fit—which would be the technological and liberalist way to sustainability that comes almost as a reflex to subjects of fossil capitalism. Rather, the scaffolding for meaningful life and the scaffolding for ecologically sustainable life are the same thing. To know how to live in a way that we would call "ecologically sustainable" and to know how to live within the parameters of meaningful human existence are hugely overlapping, if not the same thing.

3.1 *Knowledge of Bearings: Kalevala*

This type of knowledge that binds together elements of cultural meaning and ecological sustainability is explicitly named in the Karelian oral tradition that has been gathered into the epic *Kalevala*, perceived as the Finnish national epic and, maybe most famously, a crucial inspiration for the world-building by J. R. R. Tolkien. One of the most explicit presentations of the type of knowledge happens in a famous "singing contest" (poem III) between the cultural hero and sage Väinämöinen and the younger contender Joukahainen. Even though Väinämöinen gets frequently in trouble and seeks counsel from ancestors, he is presented throughout the *Kalevala* as the most knowledgeable and influential character, giving shape to the lives of others. Joukahainen, in turn, is a rash and ambitious young man, who does not listen to the advice of his parents or even of his own better self.

In the poem, Väinämöinen and Joukahainen happen to face each other on a narrow path, so that one must give way. In order to solve the situation, Joukahainen challenges Väinämöinen into a contest; either through using swords or through "singing", i.e., reciting ancient oral wisdom. Väinämöinen declines the challenge. Joukahainen provokes Väinämöinen by claiming that he, Joukahainen, knows very ancient things. This, finally, sets Väinämöinen on edge, as Väinämöinen himself was present when the world was created. The contest is on. Joukahainen starts by singing lists of, as such, correct factual knowledge of the habits of animals and geographic features of the nearby areas. Väinämöinen dismisses these songs as childish, and asks Joukahainen to sing deeper songs, "syntyjä," i.e., the "births" or "bearings" of things. Joukahainen

17 Such as architectural features with essential ecological and spiritual meaning, see Watson (2019).

does not know how to do this. It is Väinämöinen's turn, and the poem ends in a figurative section where Väinämöinen sings "the births/bearings of things." The effect of Väinämöinen's singing of "syntyjä" is that Joukahainen sinks into the mire up to his neck, his dog becomes a rock and so on. In order to save himself, Joukahainen promises his sister as a wife to Väinämöinen, and a new section in the saga begins.

Also in several other places of the *Kalevala* oral tradition, "synnyt" is mentioned. "Synty" translates crudely to "origin, cause, birth." In combination with "tieto" for "knowledge" (more literally "to know the way," "to way-find," "to make the way"), the word "syntytieto" becomes "to know what is the birth of something" or, more succinctly, "knowledge of bearings." Here two connotations of "to bear/bearings" (from the Indo-European root "bher" for "to carry a burden, bring," "give birth") are especially salient: to know one's "bearings" in terms of knowing one's place and knowing the directions and "to bear" in the sense of to suffer, to undergo, to bring and to give birth.

Typically, in the oral tradition the "knowledge of bearings" is not directly named or sung, i.e., the content of the knowledge is not made explicit in words. In the "singing contest" Väinämöinen sings and Joukahainen sinks into the mire, but what, exactly, Väinämöinen says is not spelled out. Typically, when the poems indicate that someone "knows the bearings," some kind of collective action happens, such as a man seen neck deep in a mire or a healing taking place. For a cultural hero like Väinämöinen "to know the bearings" is to be able to effect a collective change, to make people see, both in the literal and the metaphorical sense. Here, the facts listed by Joukahainen are not enough. More generally, Väinämöinen's status as the hero is precisely because and only in so far as he "knows the bearings." If it turns out that he does not, then he is not a hero anymore, and a new one may arise; someone who knows the bearings better. A hero can guide and give shape to the life of others only in so far as s/he sings "syntyjä" that actually are effective for the collective.

Thus, to "know the bearings" is to know the conditions, both immaterial and material, for the continued life of the culture: what makes sense, why are things done as they are, where are the material and immaterial bottlenecks for continued collective existence. As a very coarse characterization, we could say that knowledge of bearings combines three things, i) causal knowledge of livelihoods ("ecological sustainability"), ii) acausal knowledge of the universe of meanings ("meaningful life") and iii) knowledge of their connection (the holistic nature of any culture). To know what is the synty/bearing of this or that thing is to know what is the role of that thing (*Sache*, as well as *Dinge*) in the continuity management of that particular meaningful way of life.

When it is said that Väinämöinen was present when the world was created, the implication is that he knows the material and socio-spiritual conditions that are necessary for the continuation and development of the community. These conditions need not be inflexible. Rather, to know the bearings is to know the crucial bottlenecks that a way of life has to negotiate in order to continue and develop, and the crucial thresholds (or "planetary boundaries", as the term goes today) the crossing of which forces a mutation to a state that is recognized as different, not anymore continuous with the previous state.[18] Given that knowledge of bearings deals with the universe of meanings—what makes this particular way of life hang together and meaningful—it is understandable that it cannot be explicitly expressed in well-defined sentences, but rather can be best shown in deeds, in exemplary action.

Flexibility is needed, as both the material and immaterial conditions may change. There are some inflexible minimum conditions for physical survival (food, shelter, etc.), but after those minima are reached, what is sufficient is flexible. It might happen, for instance, that some socially or spiritually crucial practice is dependent on particular material circumstances, and then those circumstances are lost. However, a community with a robust tradition of knowing and living its bearings, may be capable of re-interpreting or re-creating the needed practice via other means. The anthropological record tells of several such cases from different continents. In the Scandinavian context, the way in which Sami peoples who used to rely on fishing, hunting, and foraging found a new sustainable subsistence in nomadic reindeer herding after being pushed away from the areas that were most conducive to fishing, hunting, and foraging, and how social, cultural, and spiritual practices were reinterpreted, offers an example.[19]

The first steps from the kinds of lists of facts presented by Joukahainen towards knowledge of bearings are taken in moments when an observation or an insight becomes binding, presents an obligation. If and when the obligation becomes a pressing duty, there is still a separation between the relevant knowledge and the person (individual); when the knowledge deepens the difference tends to disappear. Someone with "knowledge of bearings" is no longer a liberal individual, making free (subjective) choices out of independently existing

18 In the story in *Kalevala*, this mutation happens in the community for which Väinämöinen was the cultural hero: after becoming increasingly irrelevant, he is replaced by "the son of Mary" and disappears, after predicting a possible return at a later time.

19 E.g., Bergman, I., Zackrisson, O., and Liedgren, L. (2013) "From Hunting to Herding: Land Use, Ecosystem Processes, and Social Transformation among Sami AD 800–1500", *Arctic Anthropology*, 50(2).

(objective) possibilities. Rather, their personhood has been fused with wider concerns, something that is also apparent in their behavior. As Ursula K. Le Guin put it in the *Wizard of the Earth-Sea* (1968): "And the truth is that as a man's real power grows and his knowledge widens, ever the way he can follow grows narrower: until at last he chooses nothing, but does only and wholly what he must do."

In this sense, the influence, prominence or power that sages or shamans like Väinämöinen have in their communities is not the power of an individual over other individuals. Foucault has an apt simile: the king does not see the faces of all of his subjects because of some superpowers, but because the subjects have all turned their faces towards the king. The "legibility" of the faces is given to the king by turning towards him. In the same way, the "knowledge of bearings" is given to an exemplary figure through collective action that follows the "knowledge of bearings," acts as if it was true. This appears as continuous and evolving communal practices.

On the level of physical things, *Dinge*, the contrast between fossil capitalism and a culture with embedded knowledge of bearings is easy to make. A paradigmatic fossil capitalist thing, such as a plastic computer keyboard, carries minimal information of its material origin, and offers very few clues on where and how it will end up. It originates as hydrocarbons, somewhere, and ends up as microplastics, somewhere. It is on purpose made to be untraceable, replaceable, non-unique. In contrast, a thing from a culture that knows its bearings typically fuses aesthetic and functional values, carries information on its origin and materials in itself, often to the point that it is possible to identify the maker(s) from the look and feel of the thing, and the materials are easily repurposed or absorbed by the environment when the item is no longer in use. This means that these two categories of things also come with different responsibilities for the "user" in their cultural settings—the duty to learn from, maintain and preserve in the latter being much more prominent than in the former. In the extreme, some concrete things, practices or places may be recognized as sacred or holy, wholly outside the regime of replacement or substitution.

3.2 *Knowledge of Bearings: Socrates*

In order to show that a "knowledge of bearings" need not be a concoction based on marginal and disappearing cultural artifacts, we can turn to a founding figure of Western philosophy, Socrates. In this short space a proper treatment of Socrates' views is not possible. The goal is simply to point to a possible interpretation among many, one that could be presented in connection to several other historical philosophers, but one that in the case of Socrates is particularly apt as Socrates' life, i.e., his exemplary actions, have inspired not only ancient Athenians but multiple generations ever since.

One feature that has puzzled scholars for centuries is Socrates' ethical view that if someone knows what is good, then that someone will behave according to this knowledge. In other words, according to Socrates it is impossible to know what is right and still do wrong. This seems to go against the common intuition that we often find ourselves doing bad things against our better knowledge. Socrates' view is expressed, for example, in the following way: "knowledge is something noble and able to govern man, and [...] whoever learns what is good and what is bad will never be swayed by anything to act otherwise than as knowledge bids."[20] As an ethical stance, Socrates' view is sometimes labeled "ethical intellectualism,"[21] implying the idea that ethics is essentially a matter of knowledge, of cognition. To be ethical is to have correct knowledge, since right action is a necessary consequence of having the right knowledge.

However, an alternative way of interpreting what Socrates is saying is to hear him talking of "knowledge of bearings." To know the bearings of something, to know its physical and cultural origins and meanings, its contribution to the continued existence and development of a community of shared material and spiritual practices, is to act according to those bearings, so that the existence and development may continue. The emphasis in Socrates view is not on the intellectual or cognitive aspect of "to know." Rather, Socrates is pointing to a different kind of knowledge, of the type of knowledge that is noble and governs a person, a type where cognition and action are not separate.

In Socrates' Athens, this type of knowledge is most naturally combined in the idea of virtue, something attained through prolonged practice. Indeed, Socrates argues that virtue is knowledge, moreover that virtue is *a kind of* knowledge.[22] The interpretation here is simply to say that the kind of knowledge is one where there is no distinction between the cognitive content and the exemplary action—knowing a virtue is displayed in acting according to the virtue. But virtue is a kind of *knowledge* in the sense that its nature can be beneficially discussed in dialectic dialogue, even if it cannot be written down in propositions. True to form, Socrates saw the place for philosophical education in conversation, not in writing. If, in our day and age, knowledge seems,

20 Plato, *Protagoras* 352b.

21 Parry, R. and Thorsrud, R. (2021) "Ancient Ethical Theory", The Stanford Encyclopedia of Philosophy (Fall 2021 Edition), Edward N. Zalta (ed.), https://plato.stanford.edu/archives/fall2021/entries/ethics-ancient.

22 Rogers, A. (1925) "The Ethics of Socrates", *The Philosophical Review*, 34(2); Rickless, S. (1998) "Socrates' Moral Intellectualism," *Pacific Philosophical Quarterly* 79(4).

by definition, something removed or disconnected from action, so much worse for our age. There is no need for Socrates (or anybody else, for that matter) to accept such a narrow definition of "knowledge."

This kind of interpretation of Socrates' view on ethical knowledge is supported by the circumstances of his death. As is well known, his supporters and friends were surprised and even scandalized by the fact that Socrates refused to flee after receiving the death sentence. However, Socrates refusal was a coherent part of his attitude towards the court case against him. He saw that as an Athenian citizen he was bound by Athenian law. He knew the bearings of Athenian life, and thus realized that the way to uphold Athenian values—which were his values—was to be honest and fulfill the course of the court case to its end. To know what it is to be an Athenian is to act as an Athenian.[23]

From the perspective of "knowledge of bearings" the deskilled landscape of fossil capitalism looks bleak indeed. We, as citizens of fossil capitalism, "know" how to mine tantalum and cobalt from the lands of indigenous and poor people. But we do not know how to get clean drinking water for those areas, afterwards. Which, from the perspective of "knowledge of bearings" means that we do not *really* know how to mine. From the Socratic perspective, we do not know how many people the world can support, if all of them have the average European level of consumption. We do not act as if we know, which means that we do not know.

The deskilling by fossil capitalism is especially dangerous as it enters a phase where the natural environment is rapidly changing. "Knowledge of bearings" is typically produced over generations in situations where people (young and old, able and disabled, of different genders and sexualities) co-operate to produce the grounds for physical and mental well-being. Long time spans are needed, as many of the natural processes (regeneration of an area of forest, ecological succession, etc.), that form the basis of an "ecologically sustainable" and meaningful way of life are much longer than the lifespan of a human individual. Likewise, inter- and over-generational practice is also needed so that the responses to human finitude (of being born, of being mortal) can be embedded in socio cultural practices. This multi-generational practice needs stable enough conditions so that both the socio-cultural and economico-ecological

23 To be precise: Socrates knew and exemplified one particular understanding or virtue of being an Athenian, one that was somewhat puzzling yet intriguing and inspiring to his disciples and compatriots. "Socrates reveals himself not as first of all an ethical theorist aiming at a scientific definition of the moral concepts, but as a reformer of a peculiar sort" (Rogers, "The ethics of Socrates," 122).

environment give relatively predictable responses to human practices. In a situation where the natural environment behaves in extreme (compared to the historical record) ways in response to destabilizing human impacts, the development and crystallization of multi-generational "knowledge of bearings" is a difficult if not impossible task.

However, maybe Socrates' example may yet be helpful. His was a lifetime of upheaval and struggle, including war and political revolution. Yet he, within his community and circle of friends and companions, persisted in the pursuit of virtue. But we should not overlook the problems, either. It could have been possible for Socrates and his contemporaries to notice that the relationship between (Athenian) human practices (agriculture, commerce, war) and nature are in dangerous conflict. Here is one of Socrates' interlocutors, Critias describing the natural history of Attica:

> And, just as happens in small islands, what now remains compared with what then existed is like the skeleton of a sick man, all the fat and soft earth having wasted away, and only the bare framework of the land being left. But at that epoch the country was unimpaired, and for its mountains it had high arable hills, and in place of the "moorlands," as they are now called, it contained plains full of rich soil; and it had much forestland in its mountains, of which there are visible signs even to this day; for there are some mountains which now have nothing but food for bees, but they had trees no very long time ago, and the rafters from those felled there to roof the largest buildings are still sound. And besides, there were many lofty trees of cultivated species; and it produced boundless pasturage for flocks. Moreover, it was enriched by the yearly rains from Zeus, which were not lost to it, as now, by flowing from the bare land into the sea; but the soil; it had was deep, and therein it received the water, storing it up in the retentive loamy soil and by drawing off into the hollows from the heights the water that was there absorbed, it provided all the various districts with abundant supplies of springwaters and streams, whereof the shrines which still remain even now, at the spots where the fountains formerly existed, are signs which testify that our present description of the land is true.[24]

Plato lets Critias continue by describing life in the city on Acropolis and its surroundings (which "was all rich in soil"[25]). The top of Acropolis used to be inhabited by the military class, who "aimed at the mean between luxurious

24 Plato, *Critias* 111b–e.
25 Ibid., 112a.

display and meanness, and built themselves tasteful houses, wherein they and their children's children grew old and handed them on in succession unaltered to others like themselves."[26] In particular, "they watched carefully that their own numbers, of both men and women, who were neither too young nor too old to fight, should remain for all time as nearly as possible the same, namely, about 20,000."[27]

Two things stick out in Critias' story. First, the shrines at the spots of fountains. These speak to a "knowledge of bearings" that was subsequently lost. Second, the steady-state sufficiency practiced by the military class atop Acropolis. During the Golden Age of Attica, both the city and the countryside knew of the elemental connection between the material and the spiritual in forming an "ecologically sustainable" way of life, as evidenced both by conscious practices of population control and sacralization. Socrates, Critias, and Plato, as citizens of a different age, were still close enough to recognize the simultaneous familiarity and strangeness of these Golden Age practices, as presented in the oral tradition. The seeds, the examples of non-destruction and sufficiency, were there and are still here.[28]

4 Conclusion

If it is the case that the types of knowledge associated with lives that are successful in sufficiency and continuity management are systematically obscured by fossil capitalism, then an analogy from nature conservation might be helpful. For instance in the EU, a massive effort of restoration of nature areas has been put into effect.[29] The aim is to repair destroyed ecosystems and bring back nature to areas in economic use. The cost of the program is hundreds of billions of euros, and benefits are estimated in the thousands of billions. At the

26 Ibid., 112c.

27 Ibid., 112d.

28 Intriguingly, Kohei Saito (2022) argues that Marx—and Marx's ethnological sources—knew about steady-state communities and traditional ways of life from Europe and Northern America. Saito's argument goes further in proposing that Marx considered aligning the communist agenda with a sufficiency-oriented outcome rather than a progressivist one.

29 See, e.g., "Green Deal: pioneering proposals to restore Europe's nature by 2050 and halve pesticide use by 2030", https://ec.europa.eu/commission/presscorner/detail/en/ip_22_3746 and "Questions and Answers on Nature Restoration Law: restoring ecosystems for people, climate and planet" (https://ec.europa.eu/commission/presscorner/detail/en/qanda_22_3747).

same time in many EU countries relatively undisturbed natural areas are taken into economic use—for traffic, new buildings, mining, energy production, even agriculture. This discrepancy between what the right hand and the left hand are doing is jarring and irrational, as it is well known that restoration is much more expensive and resource-intensive than leaving the relatively unaffected areas in peace in the first place.[30]

So the analogy: the most effective way to encourage types of knowledge that support sufficiency is to stop practices that destroy livelihoods successful in sufficiency and continuity management and their epistemic practices. In addition to various indigenous knowledge systems, this applies also to the hundreds of millions of people who currently live within the resource boundaries of one planet while enjoying a relatively high level of social goods.

Second, sufficiency should be adopted as a way of breaking through the nafthist fog in currently over-affluent areas. Here the collective nature of knowledge of bearings should be taken seriously. Knowledge of bearings is not the possession of a single individual, and it is generated only through sustained collective effort. Levels of technology and affluence, and the amount of population are all matters available for collective control, both explicitly and implicitly. In an era of accelerating crises, the crucial elements are temporal and spatial scale: where are the common practices to be found that can sustain several generations, so that skills of continuity management can grow? This is an empirical question with no clear theoretical answers.

One thing that may give hope for a transition towards sufficiency is the diminishing role of fossil fuels in the global metabolism. The epistemological effects of oil are directly connected both to its particular material properties and to its sheer mass. Timothy Mitchell has pointed out that the birth of economics as a science focused on monetary issues benefited from the fact that fossil fuels were available in amounts that meant that "oil could be counted on not to count."[31] There was so much oil available that one needed not to worry about it, or about energy, in general. Now the worry is back: we cannot burn all the available oil and have a planet favorable for human societies. This means that finding our bearings and ending the use of fossil fuels are complementary processes. While the effects of the age of oil are nearly eternal, the fog of oil is not necessarily so.

30 Jackson, S. and Hobbs, R. (2009) "Ecological Restoration in the Light of Ecological History," *Science* 325.

31 Mitchell, T. *Carbon Democracy* (Verso, 2011), 234.

References

Altvater, E. (2007) "The Social and Natural Environment of Fossil Capitalism," *Socialist Register,* 43(1): 37–59.

Bergman, I., Zackrisson, O., and Liedgren, L. (2013) "From Hunting to Herding: Land Use, Ecosystem Processes, and Social Transformation among Sami AD 800–1500", *Arctic Anthropology,* 50(2), https://doi.org/10.3368/aa.50.2.25

Crutzen, P.J. and Stoermer, E.F. (2000) The 'Anthropocene'. *IGBP Newsletter* 41, 17–18.

Dreyfus, H. (2014) *Skillful Coping: Essays on the phenomenology of everyday perception and action*. Oxford University Press.

Ghosh, A. (2021) *The Nutmeg's Curse: Parables for a Planet in Crisis*. Chicago: University of Chicago Press.

Gibson, T. and Sillander, K., eds. (2011) *Anarchic Solidarity. Autonomy, Equality, and Fellowship in Southeast Asia*. Yale University Press.

Graeber, D. and Wengrow, D. (2021) *The Dawn of Everything*. Allen Lane.

Gray, S. (2017) "Fuelling mobility: Coal and Britain's naval power, c. 1870–1914", *Journal of Historical Geography*, 58: 92–103, https://doi.org/10.1016/j.jhg.2017.06.013.

Heidegger, M. (1954) 1977. "The Question Concerning Technology," in *The Question Concerning Technology and Other Essays*, translated by William Lovitt. Harper & Row.

Hickel, J. (2020) "The sustainable development index: Measuring the ecological efficiency of human development in the anthropocene," *Ecological Economics* 167. https://doi.org/10.1016/j.ecolecon.2019.05.011

IPCC (2018) Global warming of 1.5°C. Retrieved from https://www.ipcc.ch/site/assets/uploads/sites/2/2018/07/SR15_SPM_High_Res. Pdf

Jackson, S. and Hobbs, R. (2009) "Ecological Restoration in the Light of Ecological History". *Science* 325(5940), https://doi.org/10.1126/science.1172977

The Kalevala. (1989) Compiled by E. Lönnrot, translated by K. Bosley. Oxford University Press.

Keen, S., Ayres, R., and Standish, R. (2019) "A Note on the Role of Energy in Production". *Ecological Economics*, Volume 157: 40–46, https://doi.org/10.1016/j.ecolecon.2018.11.002.

Le Guin, U. K. (1968) *Wizard of the Earth-Sea*. Parnassus.

Malm, A. (2016) *Fossil Capital: The Rise of Steam Power and the Roots of Global Warming*. Verso.

Mitchell, T. (2011) *Carbon Democracy*. Verso.

OECD (2001) OECD Environmental Strategy for the First decade of the 21st Century. http://www.oecd.org/environment/indicators-modelling-outlooks/1863539.pdf

Parry, R. and Thorsrud, R. (2021) "Ancient Ethical Theory", The Stanford Encyclopedia of Philosophy (Fall 2021 Edition), Edward N. Zalta (ed.), https://plato.stanford.edu/archives/fall2021/entries/ethics-ancient/

Plato, Dialogues with Translations, https://www.plato-dialogues.org/links.htm

Rogers, A. K. (1925) "The Ethics of Socrates", *The Philosophical Review*, 34(2): 117–143. https://doi.org/10.2307/2179462

Rickless, S. C. (1998). "Socrates' Moral Intellectualism," *Pacific Philosophical Quarterly* 79(4): 355–367. doi:10.1111/1468-0114.00068

Rye, C. and Jackson, T. (2018) "A review of EROEI-dynamics energy-transition models," *Energy Policy*, 122: 260–272, https://doi.org/10.1016/j.enpol.2018.06.041.

Salminen, A., and Vadén, T. (2015) *Energy and Experience. An Essay in Nafthology*. MCM.

Saito, K. (2022) *Marx in the Anthropocene. Towards the Idea of Degrowth Communism*. Cambridge University Press.

Saunders, A. and Luukkanen, J. (2021) "Sustainable development in Cuba assessed with sustainability window and doughnut economy approaches." *International Journal of Sustainable Development & World Ecology* 29(2). https://doi.org/10.1080/13504509.2021.1941391

Schmelzer, M. (2016) *The Hegemony of Growth: The OECD and the Making of the Economic Growth Paradigm*. Cambridge University Press.

Smil, V. (2008) *Global Catastrophes and Trends*. MIT Press.

Steffen, W., Broadgate, W., Deutsch, L., Gaffney, O., & Ludwig, C. (2015) "The trajectory of the Anthropocene: The Great Acceleration." *The Anthropocene Review*. 2(1): 81–98. doi:10.1177/2053019614564785

Vadén, T. (2021) "What Does Fossil Energy tell us About Technology?" in P. Heikkurinen and T. Ruuska (eds.), *Sustainability Beyond Technology*. Oxford University Press.

Vadén, T., Lähde, V., Majava, A., Järvensivu, P., Toivanen, T., Hakala, E. & Eronen, J.T. (2020) "Decoupling for ecological sustainability: A categorisation and review of research literature." *Environmental Science & Policy*, 112: 236–244. https://doi.org/10.1016/j.envsci.2020.06.016

CHAPTER 8

Social Organization as Transformative Technology: from Demand-Driven Growth to Needs-Oriented Wellbeing

Karl Johan Bonnedahl

1 Introduction

Although denying the absolute biophysical limits to human expansion, the Brundtland report is still reputed to have set humanity on track towards sustainable development (SD). While it recognized major ecological challenges, it proclaimed that if we just made further efforts to manage and improve technology and social organization, this would make way for a new—sustainable—era of economic growth.[1] More than a generation later, the rhetoric seemingly prevails but the dream of sustainable development has not come true.

Advances within areas such as child mortality, access to education and gender equality in some regions of the World are overshadowed by growing ecological crises and serious economic and social inequalities. The climate crisis alone disqualifies the global development as "sustainable" equally before and after the iconic WCED report. Since the dawn of industrialisation, the atmospheric level of carbon dioxide has risen by roughly 50%, and the growth rate has increased.[2] Irreversible threshold effects, like the collapse of Antarctic and Greenland ice, with a resulting sea level rise of several metres become increasingly plausible, and the Secretary-General of the UN has described climate change as an existential threat to humanity.[3] In parallel, an alarming biological degeneration is happening, with serious consequences for ecosystem functions. Types of land and habitat are degraded or destroyed, many

1 WCED (1987). *Our Common Future.* Report of the World Commission on Environment and Development. United Nations, Overview §27.

2 NOAA (2024). National Oceanic & Atmospheric Administration. Global Monitoring Laboratory, Earth System Research Laboratories. Trends in Atmospheric Carbon Dioxide.

3 Guterres, A. (2018). Secretary-General's remarks on climate change. New York 10/9 2018. *United Nation's Secretary-General*; Armstrong McKay, et al. (2022). "Exceeding 1.5°C global warming could trigger multiple climate tipping points," *Science*, 377; Lenton, T., et al. (2019). "Climate tipping points—too risky to bet against." *Nature* 575.

 | DOI:10.1163/9789004746404_009

species populations are shrinking, and others are permanently lost—to the level of the sixth mass extinction.[4]

The fundamental cause of these and many other so called environmental problems (pollution of chemicals and plastics, eutrophication, water scarcity, et cetera) is an extraordinary expansion of human societies and activities, leading to 'fundamental shifts in the state and functioning of the Earth System'.[5] This inability of humanity to fit its activities into the planetary system, and life-threatening hazards which follow, introduced the Brundtland report, and already at the 1972 Stockholm conference it was stated that:

> A point has been reached in history when we must shape our actions throughout the world with a more prudent care for their environmental consequences.[6]

About two generations later, this has certainly not happened. Rather, with continued "development," humanity has left the for life so favourable circumstances of Holocene further behind to meet the challenges of its own making in Anthropocene.[7]

In order to change this ecologically destructive path, we need to look at sustainability and development in new ways. To begin with, not only is "sustainable development" improper for a description of most plans and actions which have taken place under such a label—if we do not just accept the two words as a label for very weak policy since the Brundtland report. Unsustainable, with its static connotation, has also become insufficient for human-nature relations as they are under continuous deterioration from earlier states. Here, António Guterres, again, made the relevant description when he opened the Stockholm + 50 conference by asking world leaders to end a "senseless and suicidal war against nature".[8] Furthermore, we should talk about *societal* rather than *environmental* problems and challenges, to better understand where causes and solutions can be found, and we should recognize that development is

4 Ceballos, G., Ehrlich, P., and Raven, P. (2020). "Vertebrates on the brink as indicators of biological annihilation and the sixth mass extinction," *PNAS*, 117(24); IPBES (2018). *The assessment report on land degradation and restoration.* Summary for policymakers.

5 Steffen, W., et al. (2015). "The trajectory of the Anthropocene: The Great Acceleration." *The Anthropocene Review* 2(1), 93; see also Krausmann et al. 2018; UNEP, 2021; Vitousek et al. 1997.

6 UN (1972). Report of the United Nations conference on the human environment. A/CONF.48/14/Rev.1. Stockholm, June 5–16, Chapter I:I, §6. See also WCED, Overview, §1.

7 Crutzen, P. (2002). "Geology of mankind: The Anthropocene," *Nature*, 415; Steffen et al. 2015.

8 UN (2022b). United Nations in Western Europe. Guterres at Stockholm + 50: "End the suicidal war against nature"; See also UNEP 2021.

dependent on ecological facts and relations. These have consequences also for societal areas of activity where environmental issues have not been lifted or even discussed before.

In this quest, the concept of sufficiency is useful. It builds on a combination of common sense (why should more, newer, and faster always be better?) and a radical challenge to destructive priorities in society which until now have been engines in development (and which will be under scrutiny later in the chapter). The acceptance of sufficiency for guidance calls for an assessment of the relevance of consumption, in contrast to the dominant view in economics and policy where individual and social aspirations are and should be unlimited. The concept contains an understanding of restrictions which can be applied to a more respectful use of nature. Thereby it brings a relevant perspective to sustainability which is useful both in terms of social justice and ecological limits, and which can guide human activities which pose risk to survival.[9]

This perspective guides the chapter which, without delving into the concept of sufficiency further, discusses how a new approach to technology, as T in the IPAT model, may provide a pathway to sustainability, and how this must be informed by alternatives to growth and demand. It begins by an examination of IPAT, with a focus on T. In subsequent sections, concepts which are critical to societal change towards sustainability are examined: concepts related to societal goals followed by concepts which represent alternative societal steering mechanisms. Finally, the alternative approach to T, in which social innovation and an increased attention to wellbeing and needs are key, is discussed.

2 IPAT

A classic explanation of socioeconomic driving forces to human pressure on the environment is given by the IPAT model.[10] It has been given different forms but a simple representation says that the environmental impact (I) is a function of how many humans there are (P, or population), how much they consume (A, or affluence) and how they do in order to arrive at that level of

9 Princen, T. (2005). *The Logic of Sufficiency*. MIT Press, 19; Spangenberg, J. and Lorek, S. (2019). "Sufficiency and consumer behaviour: from theory to policy." *Energy Policy*, 129: 1071. The application of the concept includes the areas of energy (Millward-Hopkins et al. 2020), transportation (Waygood et al. 2019) and consumption more in general (Yan and Spangenberg 2018).

10 E.g. Alcott, B. (2010). "Impact caps: why population, affluence and technology strategies should be abandoned." *Journal of Cleaner Production* 18(6); Chertow, M.R. (2001). "The IPAT equation and its variants." *Journal of Industrial Ecology* 4(4).

consumption (T, as technology). Due to interrelations between factors, it should rather be seen as I = f(P, A, T) than as I = P × A × T when it is expressed as a formula.

Looking individually at the factors on the right side in the formula, the global population (P) has increased about ten times since the beginning of the industrial revolution; a good starting point for contemplation over planetary exploitation and the role of technology (the first billion people was estimated to be reached in 1804[11]). If we instead use the 1972 Stockholm conference as point of reference, as a starting point for global high-level meetings on human-nature relations, the global P factor had doubled at the time of the follow-up conference fifty years later.[12] UN population data also show that the growth rate in per cent fell to about half between these years, but that the real increase, which is the ethically and ecologically relevant figure, was roughly the same in 2022 as in 1972. Putting a pressure on the I, these roughly 75 million new human earthlings annually have needs and aspirations that are as legitimate as they are for humans which already live in today's extremely unequal global society.

If we compare the human P with other populations, to get a glimpse of how the planet's surface and primary production is used, inequalities are clearly even bigger. As an indication, the combined biomass for humans and our domesticated animals—primarily subordinated into food production systems—has been estimated to be 35 times bigger than the biomass of the planet's wild terrestrial mammals.[13] Still, the (human) population issue as part of global policy has decreased, and statements about the environmental pressure from a growing P which were produced at the Stockholm conference and in the Brundtland report was completely absent in The 2030 Agenda. There, P remains only indirectly under reproductive health (Sustainable Development Goal no 3 and 5).

The second factor on the right side, A, should capture dimensions of human actions which are of biophysical relevance—characteristics of our lives and societies which have environmental impact. The focus is on *what* we do, whereas how we do it rather belongs to the T factor. In the classic article from 1971, Ehrlich and Holdren dealt with A and T as one factor, but also independently A is considerably more complex than P. It has alternative indicators, but a use

11 UN (1999). *The World at Six Billion*. United Nations Population Division.

12 UN (2015c). *World Population Prospects: The 2015 Revision*. DVD edition; UN (2022c).

13 Smil, V. (2011). "Harvesting the Biosphere: The Human Impact," *Population and Development Review* 37.

of consumption per capita is often reasonable.[14] Apart from its usefulness for an illustration of the vast differences in the environmental impact between individuals and categories of individuals, it relates to the importance given to economic growth in policy. This includes SD policy, which implies that A is allowed to drive I. As one key example, The 2030 Agenda promotes continuous economic growth as an integrated part of sustainable development, and it uses GDP (Gross Domestic Product) as guidance, a measure which is blind for needs, distribution and environmental consequences (Sustainable Development Goal 8.1[15]). Further, the growth system basically requires an increase in P, as new consumers, labour force and taxpayers.

Hence, not talking about P and promoting an accelerating A, the remaining factor for use in SD policy is T. This focus was also expressed by the WCED, which implicitly presented T as a composition of technology, to be understood in its customary use as connected to artefacts such as machines,[16] and social organization (see above). This composite T has continued to serve an aggregate rise in A. In other words, it has continued to materialize economic growth (as, e.g., indicated by GDP) but not *sustainable* economic growth. Obvious roles have been played by 'hard' technologies (T_H) such as mechanical or digital machines and processes, but social organization, or 'soft' technologies (T_S), such as liberal trade agreements and efficient business models, have also been instrumental in growth and development of what is understood as affluence: modern, wealthy and consumerist lifestyles and societies.[17] The relation between population and technology is more complex, but elements of T certainly enable higher human population and life expectancy, not least through new medical treatments and advances within agriculture. Technologies involved in adaptation to effects of climate change will also be increasingly crucial to keep P up. Most importantly regarding the four factors, however, and

---

14 Alcott 2010; Erlich 2014; Ehrlich and Holdren 1971. Regarding alternatives, income could be used as indicator for consumption which, in its turn, could indicate resource use and environmental impact. In general, such rough indicators would be more relevant on aggregate levels (e.g. global impact), whereas more direct measures of dimensions of affluence with environmental impact should be used in specific and local settings (like specific categories within consumption).

15 UN (2015b). *Transforming our world: The 2030 agenda for sustainable development*; for alternatives, see, e.g., Costanza et al. 2016, and Schepelmann et al. 2010.

16 Examples of areas for technology with 'direct relevance to the objectives of sustainable development' given in WCED (Chapter 3, §67) are biotechnology, genetic engineering, new energy sources, new materials, and low-waste and non-polluting technologies.

17 Cf. Alcott 2010; Ehrlich 2014. Like with all models, real examples do not always fit easily into the categories. Information technologies, as one possible case with both hard and soft features, are not explicitly dealt with in this chapter.

contrary to the Brundtland report's expectations, humanity's environmental impact (I) has risen to alarming levels.

An analysis over global material flows is illustrative. The total extraction of resources such as ore, biomass and fossil energy carriers increased twelvefold from 1900 to 2015.[18] This data has been used by the UN in its own follow-up of SDG 12 (sustainable consumption and production). It shows that human planetary exploitation, on the input side of our economies and socio-economic metabolism, accelerated during the beginning of the 21st century and increased faster than both population and economic turnover.[19] Hence, the opposite of the decoupling expected by proponents of mainstream development has happened. Another implicit and sharp critique to what development means in practice, whether or not it is informed by SD, is the extremely unequal use of these extracted resources. According to the same UN report, the class of more developed countries used an average of 26 tons of natural resources per capita during 2017; 13 times more than the class of least developed countries. This illustrates how weak sustainability, which dominates SD discourse and policy, have failed; both in terms of ecology and distribution, or, if we want, inter- and intragenerational justice.

Within this mainstream discourse, it is commonly understood that T has not yet been sufficient to counter the rising I, and hopes for progress in this respect have mainly been directed towards machine- and product-centred hard technologies. An indication already in WCED is that "technolog*" appears 285 times in the report while "social organi*", its sister concept when sustainable development is defined, only appears once outside of these two paragraphs (in relation to disruptions due to war).

Typical recent examples of the focus on T_H within SD, reflecting the exosomatic orientation of humans more in general,[20] are technologies aimed at climate change mitigation (such as new fuels and energy systems, and carbon capture and storage). As hard and soft technologies rarely exist independently, these ambitions for improvements of T_H are often supported by T_S measures, or "political, social and economic arrangements".[21] In relation to the social in social organization, however, policy efforts are generally aimed at controlling

18 Krausmann, F., et al. (2018). "From resource extraction to outflows of wastes and emissions: The socioeconomic metabolism of the global economy, 1900–2015," *Global Environmental Change* 52. See also UNEP 2024.

19 UN (2019). *The Sustainable Development Goals Report 2019*, 46.

20 Georgescu-Roegen, N. (1975). "Energy and economic myths," *Southern Economic Journal* 41(3).

21 Ehrlich, P.R. (2014). "Human impact: the ethics of I = PAT." *Ethics in Science and Environmental Politics* 14(1): 11.

environmental effects of activities rather than influencing reasons behind activities. The latter can be represented by values and priorities which result in perceived needs and certain consumption behaviours.[22] Whereas some improvements or shifts of this kind clearly contribute within the limitations of the respective systems, such as when engines or materials with high environmental impact are replaced by ones with lower impact, some crucial factors often deny overall improvements.

For one thing, the use of the new and less environmentally-detrimental engine or material is typically, as the artefact or resource it replaces, embedded in a growth system; this makes improvements in terms of environmental impact only relative to output, and thus—when absolute reductions in impact (I) are not achieved—ecologically irrelevant. This is the normal case for market-based activities, that increasing sales, of a product with environmentally upgraded features is imperative, and that such market achievements partly or fully counteract gains measured per production unit (a rebound effect.)[23] Rather than presenting gains in eco-efficiency in such cases, and presuming that *eco* would stand for ecological, a more relevant description would use economic terms, as the situation is more correctly described as one with increased productivity from natural resources. This is valid when a resource such as a material or acres of land is at the denominator and applies similarly if we chose to discuss relative achievements in relation to emissions or waste (e.g. energy supplied, or ton-miles transported, per tonne of CO_2 emitted). This would mean productivity in the use of ecosystem capacity.

To distinguish between the absolute and the relative needs to be emphasized. It is essential but often forgotten, or—with more manipulative purposes, to present and give us the idea of better product and process features—deliberately avoided by agents involved in marketing. The absolute, such as total emissions of CO_2, is the ecologically relevant information, while the relative, such as CO_2 emissions per product or mile of transportation, is economic information. It presents economic benefits from environmental costs and is what can often be improved by efficiency measures.[24]

Furthermore, a reduction in environmental impact is far from being a dominant goal for technological development. Given the relevant economic

22 Bonnedahl, K. (2021). "An Economy beyond Instrumental Rationality," in: Heikkurinen, P. and Ruuska, T. (Eds.) *Sustainability Beyond Technology*. Oxford University Press.

23 Alcott (2010); Sanne (2006).

24 Which can be seen as strictly logical as a fraction presented as pollutants/economic good can be used to describe the technology (cf. Chertow [2001], p. 2001). Increased efficiency (better technology) reduces this quotient whereas the reversed fraction, productivity in terms of the use of planetary capacity (economic good/pollutants), is increasing.

incentives, (natural) resource efficiency would be part of the calculation for such development, but a more general role of T is rather to serve a rise in A. This is the case for both T_H and T_S. It includes the exploitation, domestication and transformation of components of the non-human world (lives, species, systems, processes) which can be put into use in economic and social systems.[25] Only looking at the economically rational reasons (there are certainly also biological, psychological and social reasons behind the human drive to exploit and dominate), this outcome has a normative basis and clashes substantially with any conception of sustainability which would recognize ecological requirements and consequences of human societies: the combination of an anthropocentric world view and the market system of production and exchange of value creates an imperative to make resources—*natural capital*—out of nature, and, finally, income to the business and utility for the human consumer. Driven by demand, basically applied as a proxy for utility, elements (individuals, species, systems) in nature are put into economic and social service, and *become* natural resources as they are given human meaning and economic value in processes of *production*; a term used in a way which also denies nature the role as the real primary producer.[26]

Hence, without a profound and value-based change in human-nature relations, efforts with sustainability labels are bound to, at best, only counteract the unsustainability resulting from other and more basic priorities and operations in society. Among limitless examples we find lower emissions per mile from an expanding global pool of vehicles with massive environmental footprint for their manufacturing and infrastructure, some increase in responsible sourcing for electronic equipment with rising demand and short product life cycles, reduced energy requirement for a unit of light but an increase in the number of units, et cetera. Central normative components in this continuing unsustainable development, necessary to challenge, are growth and demand.

The ineptness of conventional environmental initiatives and the mainstream sustainable development agenda is accentuated by "lock-ins," both in existing structures (manifest in, e.g., dominant knowledge, professions and organizations) and as perceived privileges, sustained by the present system's support of (the demand for, and growth of) certain types of investment and

25 Bonnedahl, K.J. (2012) *Från ekonomiskt till hållbart, från exploatering till samexistens: En bok om att tänka om* [*From the economic to the sustainable, from exploitation to coexistence: A book about rethinking*]. Lund: Studentlitteratur.

26 In economic terms, primary production is used for activities such as farming and fishery; activities in which humans apply T to extract or harvest from nature's processes, expressed as, e.g., biological, hydrological, and geological.

consumption. Consequently, not only is direct opposition to developments which would imply lifestyle changes or devaluation of certain private and corporate assets expected; sustainability polices also conform to dominant interests and privileges in the type of solutions chosen and their level of interference (including time given for adaptation to new politically imposed circumstances). The EU-ETS has been one large scale example of this, meaning that intervention with an environmental aim should not disturb economic growth or question demand as steering mechanism.[27]

The concepts of growth and demand provide two of the most problematic obstacles to tackle if environmental impact should be reduced. They hold key roles in economic theory and policy (i.e., policies behind the unsustainable development until today), and expectations for their continued role in *sustainable* development are mainstream—not least through promoting "clean tech" innovation; through informed or conscious demand, they steer towards sustainable production. Some of this addiction can be explained by lock-ins, but there is also a need to point at the often-poor use and understanding of the concepts. This relates to, e.g., the (commonly absent) distinction between stocks and flows, between wealth and wellbeing, and, equally, between demand and needs. Further, the fact that neither growth, nor demand, contain any limitations to human consumption or features to deal with just distribution clearly argues for alternative concepts to guide sustainable development and the T of IPAT: How we do things. In relation to IPAT, growth and demand fuels the I and directs the T towards A without reference to real needs. The following sections therefore discusses problematic conventional concepts and their alternatives. First, distinctions between wealth, growth, welfare, and wellbeing are treated.

3 Concepts Related to Goals for Development

One way to look at the present overshoot, in terms of how humanity uses planetary resources, is that wealth is consumed. With this perspective, assets which are crucial for societies are focused, and not any indicator (such as GDP) which deals with *flows from* assets. 'Wealth' is an economic term which includes the assets, or capital, which can generate future income in, e.g., a region or globally.[28] Therefore, it can also be used to produce goods and

27 It is also a strong example of T_S which is expected to support innovation in T_H (cleaner processes to avoid costs for carbon credits).

28 Hamilton, K. and Hepburn, C. (2014). "Wealth," *Oxford Review of Economic Policy*, 30(1); Stiglitz et al. 2009; World Bank 2018.

services with expected benefits to people. Like with personal assets, we can use them in careful or reckless ways. If we look at present flows from assets, we can conclude that the economic turnover (growth) is too high, and, if we look at the stocks of assets, the withdrawals are too high. This contrasts with the mainstream analysis which tend to conclude that overshoot signals that T is not yet good—efficient—enough; a position which is dealt with through the chapter.

If, instead, a more long-term management was successful, the assets would be a basis for future welfare and wellbeing, which corresponds to one of the definitions of sustainability.[29] We can also change components and relations within wealth. Human, institutional, and social capital (corresponding to T_S), not least, change over time and between regions, and nature is transformed into energy sources, consumables, and physical capital (both through and to T_H).

From the perspective of weak sustainability, and the belief in substitutability, such changes are desired if the estimated value of total assets increases. This builds on ideas saying that various types of assets such as water, education and physical infrastructure can be given relevant monetary values, including estimated future returns. For to make a valuation of (what used to be) nature, the normal condition is the subordination of trees, land, pigs et cetera into economic production systems, which is a primary role for T in industrial and modern societies. These processes both drive and give arguments for economic growth: If the value is estimated to increase in a process, trees, land, and pigs *should* be transformed to products that can be sold at markets. This can also be seen as a quite remarkable combination of idealism and materialism. Values are determined subjectively (illustrated by the different uses for pigs and dogs) and translated in social processes to money, which take a role as objective units in society.

In principle, values which would not imply major environmental impact can be determined and experienced. In practice, however, economic transformation, market sales, economic growth—and A—contain significant and globally rising material components with environmental impact, both on the input and output sides.[30] Among reasons we can find that the reduction to what can be marketed—a process of commodification—prioritizes consumption over preservation, the private over the common, and material values over other types of value such as ethical and esthetical. The mountain will be exploited to become a ski resort if project owners can convince investors and relevant

29 Hamilton, K. and Hartwick, J. (2014). "Wealth and sustainability," *Oxford Review of Economic Policy* 30(1); Stiglitz et al. 2009.

30 E.g. Krausmann et al. 2018; UNEP 2024.

authorities about future profits from paying visitors. If some other people just like it as it is, it is harder for them to compete with their perception of value in today's society, and it is challenging to voice intrinsic values of the mountain or of its habitats and species.

If a more realistic and ethically sensitive approach to the production of welfare and the enabling of wellbeing is adopted, particularly for the long term, the wealth concept has relevance also in discussions on strong sustainability. This cannot be said about growth.[31] On the level of organizations, many activities can and must certainly grow in a transition process, while others must contract or disappear, but on the aggregate, and measured as GDP, growth fails to recognize many qualitative and quantitative reductions in assets that happen through economic processes. This problem, common with other monetary measures of the size of human activities (turnover, profits, income), is not only a neutral matter of calculation; the imperative to create new economic value is driving exploitation.

Further, economic growth is, at best, insensitive for distribution and unfeasible as indicator for wellbeing. Among a number of possible values in life, hence dimensions which influence wellbeing, including spiritual, esthetical and physical, the norm of economic growth overprioritizes the consumption of items and experiences which can be priced and marketed. The call for continuous relative increases in consumption also implies that satisfaction—as a clear component in wellbeing—contradicts the norm and becomes a threat to the economic system. Growth also hides differences in purchasing power (see next section) which cannot be assumed to correspond to differences in the wellbeing (or even utility) which follows from consumption; on the contrary: the marginal utility of the poor person, or 'vulnerable' to use the more common term in The 2030 Agenda, should be higher than that of the rich, or affluent, to use the IPAT term. This could lead to the consideration of whether a fall in marginal utility follows from an increase in A.[32] At the same time, material aspirations can be expected to follow economic growth upwards,

31 Costanza, R., at al. (2014). "Time to leave GDP behind," *Nature* 505.

32 This, based on an idea about objective benefits coming from the additional consumption during each new period, could call for further growth to compensate for a falling marginal utility. An alternative is to see social comparisons as the basis for growth (Hirsch, 1976), and such a social struggle could continue upwards forever. A third possible interpretation from this is that a decrease in growth should be planned to follow from a falling marginal utility (independently from the need to decrease growth due to its environmental consequences; see 'uneconomic growth,' Daly, 2014). The most affluent persons should then put a halt to their consumption first.

TABLE 8.1 Different goals for development, which can guide the T

Concept	Common meaning	Role within strong sustainability
Wealth	Assets with the capacity to generate benefits to society.	Unique values are recognized, and critical assets sustained, for future and non-human benefits.
Economic growth	Increasing monetary value of economic activity.	Focus is shifted from increasing flows to an effective and respectful management of stocks.
Welfare	Circumstances and services which cater for wellbeing in a society.	Focus on basic needs and fair distribution (sufficient provision for all).
Wellbeing	Subjective experience of the quality of life.	Recognize a broad set of qualities, including sufficiency and satisfaction, balancing the role of market transactions and personal aspirations.
Wellbeing in coexistence	Subjective quality of life in spatial and temporal contexts including non-human life and objects.	Respect for the integrity of others and recognition of intrinsic values in the non-human world.

driven by social comparison and hedonic adaptation[33] This would weaken objective components of the relation between economic progress and individual wellbeing.

To summarize, Table 8.1 suggests some distinguishing characteristics of concepts often used when development and, hence, societal goals are discussed: brief descriptions, more general than strictly from economic theory, are followed by how each concept should be approached from the perspective of strong sustainability. Although competing uses, definitions, and indicators exist for these partly floating concepts, it is important to make a distinction when, e.g., addressing resources which are available for society, the relatively objective living standards on personal level, or the perceived quality of life—circumstances which are interrelated and which can all be affected positively or negatively by activities contributing to economic growth. However, strong sustainability implies that impact on stocks (resources, assets, capital) should

33 Easterlin, R.A., et al. (2010). "The happiness—income paradox revisited," *PNAS*, 107(52).

be focused prior to the size of flows (production, output, growth) when the T, how we do things, is determined. Further, as part of wealth, which enables the production of various services related to welfare in society, and gives a base for wellbeing, critical natural assets must be protected.[34]

One important part in the achievement of this is to not only evaluate assets by reducing various asset specific values to monetary value. Such reductionism is no neutral act but affects our understanding of value and encourages us to innovate T which drive transformation, exploitation, and commodification of unique values: forests become timber, and other mammals become meat. A broader approach to the evaluation of assets, in which heterogeneity is recognized, harmonizes with a more inclusive approach to wellbeing, in which material consumption and money have less central roles.[35] The collective nature of sustainability challenges also asks us to move beyond the individual and instrumental perspectives, from which the environment is there to supply us with benefits, to a more inclusive way to recognize wellbeing, in coexistence with others and other forms of life.[36]

For too long, and in a too simplistic way, the positive sides of economic expansion have been emphasized. Obvious on the negative side is the continuous and unsustainable exploitation of the planet. More neglected is the required recurring discontent with what we have: no levels of aspiration and consumption should be sufficient. In this system, marketing and much of what is labelled innovation does not only promote what is normally connected to development, but it also cultivates the necessary *discontent with the present*: the present electronic equipment, the present means of transportation, the present clothing, et cetera. If development would instead be based on a wealth concept which respects our biophysical circumstances, and be directed towards a notion of wellbeing which allows satisfaction (i.e. sufficiency) to be a virtue and not a threat, new norms are also needed as regards how to direct activities in society. This leads to a discussion on the important difference between needs and demand, and on related concepts such as preferences, aspirations, and satisfiers.

34 Daly, H., *Beyond Growth* (Beacon Press, 1996).

35 Helne, T. and Hirvilammi, T. (2022). "Balancing needs: young unemployed Finnish adults' discourse on well-being and its relation to the sustainability transformation," *Sustainability: Science, Practice and Policy*, 18(1); Keynes 1930/1972; Max-Neef 1991. An example of a reductionist way of approaching total wealth is Dasgupta's *Economics of Biodiversity*, rooted in neoclassical economics, and giving only instrumental value to the non-human parts of the World (Treich 2022).

36 Bonnedahl 2012; Bonnedahl and Caramujo 2019.

4 Concepts Related to Mechanisms for Development

Within the SD discourse, 'needs' was a central concept early on; in the Brundtland report sometimes with the precision "essential" or "basic" added. It was relatively clear what was meant with this; something mainly related to poverty and qualitatively different from a lot of things we may feel a desire for to get, own or experience. When development is meant to meet such a notion of needs, we should also recognize that needs contain factors which decrease proportionally with development, i.e., when provisions are made so that needs step by step can be satisfied, such as when schools and health services are made accessible to people.

This should imply an adjustment of the focus of development policies. When we understand that basic needs can be largely met,[37] we should accept that satisfaction is a good state for humans and apply *sufficiency* as one guiding concept in this adjustment process. More "developed" countries would thus use other priorities than developing ones. This would connect to the old and very sound reasoning by Keynes. Exactly one hundred years ahead of the target year for The 2030 Agenda, he saw a more promising future than that of its signatories. It included different and more profound qualities in life which we could address when we had solved what he called 'the economic problem,' the struggle for subsistence.[38]

This position also harmonizes with models based on a broader set of needs of which many have less direct connection to physical and economic circumstances, and, subsequently, the provision of work opportunities, housing, or schools through traditional development initiatives. One prominent example is Max-Neef's model which includes the existential categorization into being, having, doing, and interacting.[39] Such a recognition of a broad set of needs already reduces the relative role of (having) material possessions, and the corresponding planetary pressure, but another very important component in Max-Neef's thinking, essential in transition to sustainability, is the distinction between needs and satisfiers. While fundamental needs are universal, satisfiers 'are modified according to the rhythm of history and vary according to culture and circumstance'.[40] These can also be of a negative kind, with paternalism as

37 They will never be fully met, and some new basic needs of this kind also appear, due to, e.g., new diseases, war, and environmental disruptions.

38 Keynes 1930/1972.

39 Max-Neef, M. (1991). *Human Scale Development*. Apex Press. See Helne and Hirvilammi 2022 for a similar model.

40 Max-Neef (1991), 28.

an example of an inhibiting satisfier for the need of protection, or economic indicators as pseudo-satisfier for the need of understanding.

Another position on development, however, is what characterizes modern societies. It is to motivate continuously new aspirations which replace the fundamental needs and the aspirations which are met. A major problem with this position can be illustrated by the following passage from the Brundtland report:

> The essential needs of vast numbers of people in developing countries for food, clothing, shelter, jobs—are not being met, and beyond their basic needs these people have legitimate aspirations for an improved quality of life.[41]

The quotation first addresses essential needs, but the problem arrives if no limits are given for *legitimate aspirations*. Nor can such limits be allowed if eternal growth (per capita) should be a goal in SD, as it is within conventional development. While the Brundtland report admitted that perceived needs were determined by social and cultural circumstances, and that a living standard 'beyond the basic minimum' required that the consumption was 'within the bounds of the ecological possible',[42] this has not been seriously followed up in global SD policy. The problem is not dealt with by The 2030 Agenda, which only keeps some examples of what clearly correspond to the essential needs of WCED (including *basic services*, SDG 1.4, and *essential medicines*, SDG 3.8). Part of the explanation can be the technical difficulty to define needs, but, more importantly, a discussion about basic needs presupposes a position on what is *not* basic and necessary. This would pose serious threats to priorities, positions, and investment in today's society. Growth targets could no longer be applied without discrimination between more and less important activities, specific firms and industries would need to be questioned and the direction and governance of societies would need to be changed fundamentally. The states behind The 2030 Agenda were not prepared to do this. Instead, categories such as *vulnerable* and *poverty* are used. Many targets (under the 17 SDGs) still deal with basic needs (see 2.1, 6.1 and 11.1, together with the examples above), but others deal with broader social interests such as equality and non-discrimination, ambitions concerning infrastructure and cities, and growth and industrialization.

The Agenda does not propose limits for human aspirations, which would apply sufficiency to the A in IPAT. While it targets *sustainable patterns of*

41 WCED 1987, Chapter 2, §4.

42 Ibid., Chapter 2, §5.

consumption (SDG 12), it does not problematize *levels* of consumption or target the chase for social positions with means which are environmentally detrimental (unsustainable satisfiers and technologies). It does not say that the World's most affluent, despite immense inequalities in terms of both economic possibilities to satisfy basic needs and per capita environmental impact,[43] should reduce their consumption. This absence, together with the explicit call for continued economic growth, illustrates how The Agenda adds to, but does not change, the existing course of development. Its call for a continued increase in the production of economic goods and services, even if some success in a sustainable branding of these is achieved, corresponds to a continued use of T for to supply satisfiers of kinds which dominate modern society and where satisfaction of demand is in focus; not satisfaction of needs.

In the mainstream application of the concept, demand, often without definition or problematization, is used as a signal for the need of something (where also 'need' remains unqualified), to guide the application of T to use resources and direct growth. Such a relation between demand and needs is, however, not accurate, and if we would have the ambition to simultaneously address basic needs and moderation, demand as steering mechanism is an obstacle. While demand is normally assumed to correspond to utility, it emerges from contextualized preferences. In the typical modern, developed, context, social comparison and ideas of steady increases and renewal play important roles—implying dimensions of both space (the social) and time (the increase). To the determining context, systemic factors such as the infrastructure and supply of goods also belong. Further, demand is conditioned by purchasing power.[44]

Where essential needs of the kind addressed by WCED are largest, and where The 2030 Agenda uses vulnerable as concept, purchasing power is normally very limited. To guide society with demand therefore means that the development and use of technologies, and its application to resources and growth, are to the benefit of the most affluent. To change this situation and approach strong sustainability, one step is to understand and apply differences between concepts in table 2; to change the present role of demand and find better steering mechanisms.

43 See e.g. Alvaredo et al. 2018; Chancel and Piketty 2015.

44 This means that we deal with the most common use of the concept, demand for goods and services on markets. Translated to other situations, the (non-monetary) power to present claims about preferences also vary widely between groups and individuals. Demand on labor markets would imply a different analysis.

TABLE 8.2 Different mechanisms for development, which can guide the T

Concept	Common meaning	Role within strong sustainability
Demand	Market signal based on purchasing power and preferences related to social context and supply.	Steering mechanism for private consumption after equalization of purchasing power (via adjusted salaries and taxes) and a supply adapted to social justice and environmental carrying capacity.
Preferences	An individual's conscious or actual interest for something (good, service, experience).	Achievement of more sustainable preferences through new norms and structures.
Needs	More fundamental than preferences, such as physical, emotional or social needs.	Needs prioritized over demand and particular satisfiers.
Sufficiency in coexistence	Focus on what is enough for yourself, while respecting the same principle being applied for others, including non-human life and objects.	Highlighting common dimensions in development and non-instrumental relations to others and nature.

5 Social Organization as Transformative Technology

As argued, the mainstream approach to SD, today largely represented by The 2030 Agenda, puts faith into adjustments of parts of one (T) of the three variables which determine the theoretical I of IPAT and the actual increases in global ecological imbalances. This approach of weak sustainability builds on assumptions of high substitutability between various forms of assets, or capital. The call for (eternal) economic growth presupposes that total gains can be reached by continuous transformation of natural assets into corporate profits, buildings, infrastructure et cetera. Exploitation is believed to turn into something careful, smart, efficient, and sustainable. In policy, this is often phrased as a presumed self-evident and necessary balance and integration between economic, social, and environmental dimensions.[45] In practice, this balance does

45 Expressed in §2 of The 2030 Agenda as: "We are committed to achieving sustainable development in its three dimensions—economic, social and environmental—in a balanced and integrated manner." (UN 2015b). See also UN 2002.

not refer to ecological balance; balance between human societies and their biophysical bases. It refers to trade-offs (and substitution) between values in the different dimensions. The call for growth in one of these—the economic dimension—contradicts any real ambitions for balance between society and nature.

This is ecologically ignorant but politically logical. The dominant SD policy is a result of social and economic concerns, and of political compromise. Isolated in time, space and dominant structures, the outcome, such as The 2030 Agenda, may be seen as reasonable. It does deal with aspects of all three dimensions which are declared to be important. It would also be more surprising if present privileges and power structures—including how they relate to coming human generations and to the non-human world—would not play out in programmes for development and in decisions concerning the use and distribution of resources.[46]

If, however, SD policy should make ecological sense, and therefore also be sensible for other species and future generations, a different and more ambitious approach is needed. One alternative would be to shift focus within the T factor. Today's weak T_S largely aims to support developments of T_H, and, without disturbing markets and the individual choice, to influence production and consumption patterns to lower the impact of IPAT's A (this is typical for innovation policy, and a key feature of *The European Green Deal*). A strict top-down alternative could build on components such as absolute limits for the use of critical resources, including the use of nature to absorb emissions and waste. Combinations of taxes, quotas and caps, and protection of land and waters, considerably stronger than today, would be key parts of such policy.

Another but related alternative, which is in focus in this section, is social innovation. It does not exclude top-down measures but rather depend on institutional changes to support various often local and bottom-up initiatives for change in social practice.[47] Similar approaches have also been discussed as (the need for) cultural evolution, including the change of values that govern much of human activity, and as the development of cultural capital, "the wide variety of ways in which societies interact with their environment."[48] Moreover, Max-Neef

46 Cf. Gardiner, S.M. (2006). "A Perfect Moral Storm: Climate Change, Intergenerational Ethics and the Problem of Moral Corruption." *Environmental Values* 15(3).

47 E.g. Grimm, R., Fox, C., Baines, S. et al. (2013). "Social innovation, an answer to contemporary societal challenges?" *Innovation: The European Journal of Social Science Research* 26(4); Howaldt et al. 2014; Spangenberg and Lorek 2019.

48 Quote from Berkes, F. and Folke, C. (1992). "A systems perspective on the interrelations between natural, human-made and cultural capital." *Ecological Economics* 5(1), 2. For cultural evolution, see Ehrlich (2014).

discusses cultural change as involving change in the type and quantity of satisfiers—while the fundamental needs remain the same.[49] In coherence with these authors, Yan and Spangenberg define sufficiency, in a discussion on a needed change of consumer aspirations, as:

> a state of human living characterized by the fulfilment of needs in harmony with society and environment through the ethically justified choice of sustainable satisfiers, resulting in human well-being while safeguarding the integrity of the social and natural environment.[50]

Potentially, social innovation deals with such change, which have the *how* we do things, and T of IPAT, at its core. While it traditionally has been applied to social issues in isolation from biophysical basics, the growing ecological crises make it increasingly more relevant to also apply social innovation to relations between society and the environment.[51] However, the limitation in its prior application calls for some caution, and also for the development of approaches to social innovation which could solve and not reproduce ecological problems. Whereas social innovation research tends to be very normative and take much of the socially good as given,[52] we need to consider that the socially desirable is a matter of social negotiation and power. Hence, the socially innovative is not necessarily "good" per se or leading to fair or sustainable states. A lot of socially innovative things have happened under horrible flags, such as the one of Nazi Germany.

However, some components of social innovation have promising features. Definitions often depart from unmet needs and other challenges faced by society and focus on ways to increase the capacity to address needs and challenges through, e.g., new forms of cooperation.[53] This, to take both the goals and means into consideration, and not only focus on efficiency within present systems is a crucial part of transition to sustainability. As the ecological

49 Max-Neef, *Human Scale Development*, 18.

50 Yan, B. and Spangenberg, J.H. (2018). "Needs, wants and values in China: reducing physical wants for sustainable consumption." *Sustainable Development* 26, 778. See also 'Sufficiency in coexistence', in the second table above.

51 Haskell, L., Bonnedahl, K.J. and Stål, H. (2021). "Social innovation related to ecological crises: A systematic literature review and a research agenda for strong sustainability." *Journal of Cleaner Production* 325; Olsson et al. (2017).

52 Segnestam Larsson, O. and Brandsen, T. (2016). "The Implicit Normative Assumptions of Social Innovation Research—Embracing the Dark Side," in Brandsen, T., et al. (eds.). *Social Innovations in the Urban Context*. Springer.

53 Bonnedahl et al. (2022); Grimm et al. (2013); Schartinger et al. (2020).

challenges are caused by conventional development, guided by concepts such as economic growth and demand, not only capacity but new ways to look at development and its goals and means are needed. Here, social innovation could contribute with a shift of focus from the individual to the collective, from the more atomistic market perspective where private benefit is in focus to cross-sectoral solutions of common problems.[54] This would be both in terms of aim and means, including new types of organizations with non-commercial purposes. We would, hence, see more of the social T and less of the exosomatic one in Georgescu-Roegen's (1975) meaning.

Subsequently, key for understanding change from the perspective of social innovation are social practices, which refer to common features within social settings and relate to actions as well as to objects, meaning and knowledge which are associated to actions.[55] A related multidimensional change is therefore needed, particularly in areas with the most direct importance for ecological imbalances. These can be framed as critical practice fields, such as transportation, energy production and tourism.[56] Here, real examples of social innovation are often relatively local, emphasising factors such as civic empowerment and the wellbeing of communities. They show that social innovation in various ways can imply change in practices—part of new social organization, and transformative T_S—which restore and build aspects of wealth both in terms of social and natural capital.

Clearly, this is more a potential than a reality today. What must be included in social innovation processes, related to the dimension of knowledge above, is that economic throughput and the following accumulation of man-made capital and its wastes must decrease. This is another way of formulating that drivers of I must be understood properly to be addressed properly. Hence, the development of knowledge should be more profound and disruptive than such needed for a continued focus on T_H—aiming to only reduce the relative environmental impact following from priorities and actions in modern society.

The abandonment of growth poses particular problems; both in terms of growth as rise in A and in terms of the role given to T to make that rise happen. Obviously, but yet to be acknowledged by mainstream policymakers,

54 Ziegler, R. (2017). "Social innovation as a collaborative concept." *Innovation: The European Journal of Social Science Research* 30(4).

55 Howaldt et al. (2014). See also Spangenberg and Lorek (2019).

56 Schartinger, D., Wepner, B., Andersson, T. et al. (2017). *Social Innovation in Environment and Climate Change: Case study results*. SI-DRIVE D6.3. Dortmund Technische Universität. See also Batle et al. (2018); Hoppe and De Vries (2019); Jaeger-Erben et al. (2015).

business school educators and others, continuous growth in throughput on a finite planet is impossible in ecological terms. Still, the continued persuasion in sustainability policy that economic growth, and in principle an eternally rising A, is central to development may not primarily be explained by a clash of knowledge paradigms (e.g. on real limits to substitutability, capacities of ecosystems or availability of land and water). More likely, it is a matter of ethics and politics. Regarding ethics, present decision-makers are inclined to prioritize increases in their (our) own affluence over a fair distribution between humans and species, or over a preservation of wealth for future generations. Concerning politics, an abandonment of growth as policy goal would seriously impede the continuous short-term distribution of economic benefits, or hopes for such distribution, to various social groups and actors, which may be necessary for social stability. This would refer both to stability at the bottom of socio-economic pyramids (to avoid, in the worst case, uprising and revolution) and to stability within the higher levels, where most of the benefits from planetary exploitation and economic activity are allocated.

A change of this is challenging, to say the least. Nevertheless, apart from the long-term necessity given by ecological realities, change has promising short-term social and ethical consequences. To achieve these, a more sophisticated understanding must replace the simplified and often flawed connection made between increases in production and consumption, on the one hand, and wealth and wellbeing on the other. Today, much of the societal efforts are directed by market demand, preferences made manifest through economic capacity, to indicate wellbeing, via (growing) consumption. To replace this ecologically detrimental relation between growth and demand thus also means to move away from poor indicators, to arrive closer to what they are assumed to indicate (see Tables above). Here, governance and allocation should target needs in order to promote wellbeing and do so in an ethically inclusive and ecologically aware context so that wellbeing in coexistence becomes an overarching societal goal.

This means that the needs concept must be used in ways which distinguish between the more and less essential. To drive a car, eat meat or upgrade to the latest smartphone today are no needs. These are examples of possible context-dependent preferences, directed to different satisfiers, and enabled by social priorities and economic resources. Such preferences and satisfiers can be changed in the social and cultural development which is a necessary part of transition to sustainability. Needs should also be understood in an inclusive way, countering the excess for some and injustice in relation to others, meaning that sufficiency and coexistence also will guide the use of resources. Hence, moderation need to appear alongside consumption in our practices, and stewardship alongside production.[57] To relate to the dimension of meaning above, the socially innovative should often be to slow down, relax, and avoid, and T must be applied with reconnection to nature as a goal, in contrast to the distancing built into the T of modern societies, particularly T_H.[58]

To make the move from demand to needs more than semantic is a serious challenge. Real needs have to get voice; to be understood and respected separately from the dominant perspectives, privileges and socio-economic positions of power. Also in a room where social innovation is discussed, parties which are present have interests which are bound to the time and place (me, now, here). These spatial and temporal characteristics of unsustainability also create a particular structural problem: how could nature and its species, the most underprivileged humans, and interests distant in space and time be heard? Could it be so that the intentions and capacities of humanity to achieve sustainability are overrated? The question is at least motivated by assessing the talk versus the walk. Humanity may indeed not be ethically or biologically suited for prioritizing sustainability. Maybe output from the discourse of sustainable development should be seen as social and cultural expressions among others, or even as means for moral comfort on our destructive path, and not like serious action plans.[59]

Even if this would be the case, it does not absolve you or me from responsibility. Independent of what position we take on issues such as individual consumption or societal goals, we are moral agents. That is what we are in a world with others, and where much of the perceived pleasure, through experiences and activities connected to affluence, is available with the costs of exploitation and injustice, and finally the destruction of habitats and extinction of species. It is an intellectual and moral obligation to recognize these connections. As the chapter has shown, however, these connections are obscured by the imperative roles of demand and growth; concepts which are misused in a mainstream context and inaccurate as guides if we aim at sustainability.

Nonetheless, change is not only dependent on knowledge. It would require cultural evolution with development in terms of both ethics and systems of governance, hence T_S and not T_H. Some inspiration to how we can approach

57 cf. Dietz, R. and O'Neil, D. (2013) *Enough is Enough*. Berrett-Koehler Publishers; Princen (2005).

58 Bonnedahl, K.J. and Caramujo, J. (2019) "Beyond an Absolving Role for Sustainable Development: Assessing Consumption as a Basis for Sustainable Societies." *Sustainable Development* 27(1); Georgescu-Roegen (1975); Olsson et al. (2017).

59 Bonnedahl and Caramujo (2019); Daly (1999); Gardiner (2006).

this by including the other in our understandings and decision-making can be found in the ideas of John Rawls and his famous veil of ignorance.[60] A more hands-on approach, suitable for social innovation initiatives, is to include change-oriented organizations and networks from civil society in decision processes, but this would also need support by institutional reform, not least in business and economics education, where change in the use of terminology as shown is needed. Clearly, this should also inform policy, including, but not restricted to, the abandonment of economic growth as policy goal. Among other changes needed in the policy field are updates of legislation and trade agreements to compensate for the superior position held by economic priorities in relation to ideas of fairness and values represented in nature.[61]

In relation to the scaling or diffusion of relevant social innovations, this can illustrate that it is not enough to only scale out, or disseminate. The type and magnitude of change needed also calls for scaling up, to change institutions, and scaling deep, to impact on values and beliefs.[62] This would also put conflicts on the table. Both SD and social innovation are too often presented as agendas or processes towards common goals in harmony. This is often false, and not least when multiple parties discuss development, caution should be taken if something appears to be a win-win situation. Radical change will per definition lead to losses for certain positions and privileges, particularly strong ones, and such losses must also be a development goal.

Acknowledgements

This chapter is a substantially extended and reworked version of a chapter in Swedish, with the permission of the publisher Studentlitteratur, Lund. The full reference is Bonnedahl, K.J. (2022) *Från efterfrågedriven tillväxt till behovsdrivet välbefinnande. En roll för social innovation?* Pp. 167–184 in Bonnedahl, K.J., Sjölander, A.E. and Lindberg, M. (2022), *Social innovation för hållbar utveckling*. Lund: Studentlitteratur.

60 Rawls, J. (1971/1999). *A Theory of Justice*. Revised edition. Harvard University Press. In this context it would ask us to forget our current position in the World (rich or poor, in good health or not, etc.) before we promote certain social order or change.

61 Cf. Ezra (2017).

62 Moore and Riddell (2015).

References

Alcott, B. (2010). "Impact caps: why population, affluence and technology strategies should be abandoned." *Journal of Cleaner Production* 18(6): 552–560.

Alvaredo, F., Chancel, L., Piketty, T. et al. (2018). *World Inequality Report 2018*. World Inequality Lab. https://wir2018.wid.world/ Accessed April 24, 2019.

Armstrong McKay, D.I., Staal, A., Abrams, J.F. et al. (2022). "Exceeding 1.5°C global warming could trigger multiple climate tipping points." *Science*, 377(6611).

Batle, J., Orfila-Sintea, F., Moon, C.J. (2018). "Environmental management best practices: Towards social innovation." *International Journal of Hospitality Management* 69: 14–20.

Berkes, F. and Folke, C. (1992). "A systems perspective on the interrelations between natural, human-made and cultural capital." *Ecological Economics* 5(1): 1–8.

Bonnedahl, K.J. (2012) *Från ekonomiskt till hållbart, från exploatering till samexistens: En bok om att tänka om* [*From the economic to the sustainable, from exploitation to coexistence: A book about rethinking*]. Studentlitteratur.

Bonnedahl, K.J. (2021). "An Economy beyond Instrumental Rationality," in: Heikkurinen, P. and Ruuska, T. (eds.) *Sustainability Beyond Technology*. Oxford University Press.

Bonnedahl, K.J. and Caramujo, J. (2019) "Beyond an Absolving Role for Sustainable Development: Assessing Consumption as a Basis for Sustainable Societies." *Sustainable Development* 27(1): 61–68.

Bonnedahl, K.J. and Heikkurinen, P. (eds.) (2019). *Strongly Sustainable Societies*. Routledge.

Bonnedahl, K.J., Sjölander, A.E. and Lindberg, M. (2022). *Social innovation för hållbar utveckling* [*Social innovation for sustainable development*]. Studentlitteratur.

Ceballos, G., Ehrlich, P.R., and Raven, P.H. (2020). "Vertebrates on the brink as indicators of biological annihilation and the sixth mass extinction." *PNAS*, 117(24): 13596–13602.

Chancel, L. and Piketty T. (2015). *Carbon and Inequality: From Kyoto to Paris*. Paris School of Economics.

Chertow, M. (2001). "The IPAT equation and its variants." *Journal of Industrial Ecology* 4(4): 13–29.

Costanza, R., Kubiszewski, I., Giovannini, E. et al. (2014). "Time to leave GDP behind." *Nature* 505: 283–285.

Costanza, R., Daly, L., Fioramonti, L., et al. (2016). "Modelling and measuring sustainable wellbeing in connection with the UN Sustainable Development Goals." *Ecological Economics* 130: 350–355.

Crutzen, P.J. (2002). "Geology of mankind: The Anthropocene." *Nature*, 415: 23.

Daly, H. E. (1996). *Beyond Growth*. Beacon Press.

Daly, H. E. (1999). "Free trade, sustainable development and growth: some serious contradictions—a review," in H. E. Daly (ed.), *Ecological Economics and the Ecology of Economics*. Edward Elgar.

Daly, H. E. (2005). "Economics in a full world." *Scientific American*, 293(3): 100–107.

Daly, H. E. (2014). *From Uneconomic Growth to a Steady-State Economy*. Edward Elgar.

Dietz, R. and O'Neil, D. (2013) *Enough is Enough*. Berrett-Koehler Publishers.

Easterlin, R.A., McVey, L.A., Switek, M. et al. (2010). "The happiness – income paradox revisited." *PNAS*, 107(52): 22463–22468.

Ehrlich, P.R. (2014). "Human impact: the ethics of I = PAT." *Ethics in Science and Environmental Politics* 14(1): 11–18.

Ehrlich, P.R. and Holdren, J.P. (1971). "Impact of Population Growth." *Science*, 171: 1212–1217.

Ezra, O. (2017). "The Rights of Non-Humans: From Animals to Silent Nature." *Law & Ethics of Human Rights*, 11(2): 285–304.

Gardiner, S.M. (2006). "A Perfect Moral Storm: Climate Change, Intergenerational Ethics and the Problem of Moral Corruption." *Environmental Values* 15(3): 397–413.

Georgescu-Roegen, N. (1975). "Energy and economic myths." *Southern Economic Journal* 41(3): 347–381.

Grimm, R., Fox, C., Baines, S. et al. (2013). "Social innovation, an answer to contemporary societal challenges?" *Innovation: The European Journal of Social Science Research* 26(4): 436–455.

Guterres, A. (2018). Secretary-General's remarks on climate change. New York 10/9 2018. *United Nation's Secretary-General*. https://www.un.org/sg/en/content/sg/statement/2018-09-10/ Accessed Oct 10 2018.

Hamilton, K. and Hartwick, J. (2014). "Wealth and sustainability." *Oxford Review of Economic Policy* 30(1): 170–187.

Hamilton, K. and Hepburn, C. (2014). "Wealth." *Oxford Review of Economic Policy*, 30(1): 1–20.

Haskell, L., Bonnedahl, K.J. and Stål, H. (2021). "Social innovation related to ecological crises: A systematic literature review and a research agenda for strong sustainability." *Journal of Cleaner Production*, 325, 129316.

Helne, T. and Hirvilammi, T. (2022). "Balancing needs: young unemployed Finnish adults' discourse on well-being and its relation to the sustainability transformation." *Sustainability: Science, Practice and Policy*, 18(1): 158–170.

Hirsch, F. (1976). *Social Limits to Growth*. Harvard University Press.

Hoppe, T. and de Vries, G. (2019). "Social Innovation and the Energy Transition." *Sustainability* 11, 141.

Howaldt, J., Butzin, A., Domanski, D. et al. (2014). *Theoretical Approaches to Social Innovation – A Critical Literature Review*. SI-DRIVE. Sozialforschungsstelle.

IPBES (2018). *The assessment report on land degradation and restoration.* Summary for policymakers. Intergovernmental Science-Policy Platform on Biodiversity and Ecosystem Services.

Jaeger-Erben, M., Rückert-John, J. and Schäfer, M. (2015). "Sustainable consumption through social innovation: a typology of innovations for sustainable consumption practices." *Journal of Cleaner Production* 108: 784–798.

Keynes, J. M. (1930/1972). "Economic Possibilities for Our Grandchildren," in Keynes, J. M., *The Collected Writings of John Maynard Keynes*. Vol. IX. Macmillan Press.

Krausmann, F., Lauk, C., Haas, W. et al. (2018). "From resource extraction to outflows of wastes and emissions: The socioeconomic metabolism of the global economy, 1900–2015." *Global Environmental Change* 52: 131–140.

Lenton, T., Rockström, J., Gaffney, O. et al. (2019). "Climate tipping points — too risky to bet against." *Nature* 575: 592–596.

Max-Neef, M. (1991). *Human Scale Development.* The Apex Press.

Millward-Hopkins, J., Steinberger, J.K., Rao, N.D. et al. (2020). "Providing decent living with minimum energy: A global scenario." *Global Environmental Change*, 65, 102168.

Moore, M-L. and Riddell, D.J. (2015). "Scaling out, Scaling up, Scaling deep: Advancing systemic social innovation and the learning processes to support it." *Technical Report in Journal of Corporate Citizenship,* June 2015.

NOAA (2024). National Oceanic & Atmospheric Administration. Global Monitoring Laboratory, Earth System Research Laboratories. Trends in Atmospheric Carbon Dioxide. https://gml.noaa.gov/ccgg/trends/gr.html. Accessed June 17 2024.

Olsson, P., Moore, M.L., Westley, F.R. et al. (2017). "The concept of the Anthropocene as a game-changer: a new context for social innovation and transformations to sustainability." *Ecology and Society* 22(2): 31.

Princen, T. (2005). *The Logic of Sufficiency.* MIT Press.

Rawls, J. (1971/1999). *A Theory of Justice.* Revised edition. Harvard University Press.

Sanne, C. (2000). *Rekyleffekten och effektivitetsfällan – att jaga sin egen svans i miljöpolitiken.* Rapport 5323. Naturvårdsverket.

Schartinger, D., Wepner, B., Andersson, T. et al. (2017). *Social innovation in environment and climate change: Case study results.* SI-DRIVE D6.3. Dortmund Technische Universität.

Schartinger, D., Rehfeld, D., Weber, M. et al. (2020). "Green social innovation – towards a typology." *European Planning Studies*, 28(5): 1026–1045.

Schepelmann, P., Goossens, Y. and Makipaa, A. (eds.) (2010). *Towards Sustainable Development: Alternatives to GDP for measuring progress.* Wuppertal Institute for Climate, Environment and Energy.

Segnestam Larsson, O. and Brandsen, T. (2016). "The Implicit Normative Assumptions of Social Innovation Research—Embracing the Dark Side," in Brandsen, T., Cattacin, S., Evers, A. & Zimmer, A. (eds.). *Social Innovations in the Urban Context.* Springer.

Smil, V. (2011). "Harvesting the Biosphere: The Human Impact." *Population and Development Review* 37: 613–636.

Spangenberg, J.H. and Lorek, S. (2019). "Sufficiency and consumer behaviour: from theory to policy." *Energy Policy*, 129: 1070-1079.

Steffen, W., Broadgate, W., Deutsch, L. et al. (2015). "The trajectory of the Anthropocene: The Great Acceleration." *The Anthropocene Review* 2(1): 81–98.

Stiglitz, J.E., Sen, A. and Fitoussi, J-P. (2009). *Report by the commission on the measurement of economic performance and social progress.* https://www.economie.gouv.fr/files/finances/presse/dossiers_de_presse/090914mesure_perf_eco_progres_social/synthese_ang.pdf Accessed July 9, 2021.

Treich, N. (2022). "The Dasgupta Review and the Problem of Anthropocentrism." *Environmental and Resource Economics*, 83: 973–997.

UN (1972). Report of the United Nations conference on the human environment. A/CONF.48/14/Rev.1. Stockholm, June 5–16, 1972. United Nations.

UN (1999). *The World at Six Billion.* United Nations Population Division.

UN (2002). *Johannesburg Declaration on Sustainable Development.* World Summit on Sustainable Development. United Nations.

UN (2015a). *The Millennium Development Goals Report 2015.* United Nations.

UN (2015b). *Transforming Our World: The 2030 agenda for sustainable development.* United Nations.

UN (2015c). World Population Prospects: The 2015 Revision. DVD edition. United Nations (Department of Economic and Social Affairs, Population Division). http://esa.un.org/unpd/wpp/Download/ Standard/Population. Accessed April 22, 2017.

UN (2019). *The Sustainable Development Goals Report 2019.* United Nations.

UN (2022a). *The Sustainable Development Goals Report 2022.* United Nations.

UN (2022b). United Nations in Western Europe. Guterres at Stockholm + 50: "End the suicidal war against nature". https://unric.org/en/guterres-at-stockholm50-end-the-suicidal-war-against-nature/ Accessed July 24, 2022.

UN (2022c). United Nations, Department of Economic and Social Affairs, Population Division. World Population Prospects 2022. Summary of results. https://www.un.org/development/desa/pd/sites/www.un.org.development.desa.pd/files/wpp2022_summary_of_results.pdf. Accessed July 25, 2022.

UNEP (2021). *Making Peace with Nature: A scientific blueprint to tackle the climate, biodiversity and pollution emergencies.* United Nations Environment Programme.

UNEP (2024). *Global Resources Outlook 2024: Bend the Trend—Pathways to a liveable planet as resource use spikes.* International Resource Panel. United Nations Environment Programme.

Vitousek, P.M., Mooney, H.A., Lubchenco, J. et al. (1997). "Human Domination of Earth's Ecosystems." *Science*, 277(5325): 494–499.

Waygood, E.O.D., Sun, Y. and Schmöcker, J-D. (2019). "Transport sufficiency: Introduction and case study." *Travel Behaviour and Society*, 15: 54–62.

WCED (1987). *Our Common Future.* Report of the World Commission on Environment and Development. United Nations.

World Bank (2018). *The Changing Wealth of Nations 2018.* The World Bank Group.

World Bank (2022) Data /GDP (current USD) https://data.worldbank.org/indicator/NY.GDP.MKTP.CD?view=chart Accessed July 29, 2022.

Yan, B. and Spangenberg, J.H. (2018). "Needs, wants and values in China: reducing physical wants for sustainable consumption." *Sustainable Development*, 26: 772–780.

Ziegler, R. (2017). "Social innovation as a collaborative concept." *Innovation: The European Journal of Social Science Research* 30(4): 388–405.

CHAPTER 9

Human-Powered Transit, Energy, and the Sufficiency of Slow Travel

J. Mohorčich

1 Introduction

Moving people and things is constrained by the energy available to move people and things. Walking, for example, requires calories in the form of food. Driving requires calories in the form of gasoline. Given this fact, this chapter considers the feasibility of six modes of transportation under two scenarios: one in which humans have 50% less energy to use for transportation and one in which we have 95% less energy for transportation. I analyze the effects transportation mode choice and speed have on energy consumption. Mode and speed, I argue, are likely to limit the physical size of human settlements.

The current organization of life on Earth is dominated by the fact that human beings, largely because of their exclusive access to underground fossil fuel reserves and their domination of the Earth's surface area, physical resources, and other animals, direct a large amount of the energy flows on the planet. Total human energy production can be modeled using a version of the Impact = Population * Affluence * Technology formula, or IPAT. Total human energy production (impact, in this example) can be expressed by multiplying human population (P) by the demand for energy per person (affluence, in the IPAT model) and technology (T), which can act as a moderating or accelerating factor on energy demanded. (Newly-efficient technologies can decrease energy demanded per unit of affluence, but power-hungry technologies can increase it.) The primary energy supply of all human societies is about 585 exajoules per year. (In IPAT terms, the math is roughly 8 billion people * ~73 GJ per person = 585 exajoules. Note that in reality energy per person is hugely unequal, with some societies allocating more energy to their pets than humans in other parts of the world use.) A few hundred exajoules is minuscule compared to the amount of energy in the form of solar radiation that reaches the Earth (about 2.8 million exajoules per year), but enormous compared to what other organisms, including past humans, have been able to use.

But: people living after the era of human dominance will likely control a smaller fraction of the Earth's surface area and material resources than they

 | DOI:10.1163/9789004746404_010

do today.[1] As a result, these humans will have less control over the energy flows and stocks on and in Earth. They will have less energy available to move and make things. The exigencies of climate change, resource depletion, the difficulty and mounting inefficiency of extracting harder-to-reach reserves, breakdowns in the large-scale organization needed for energy extraction, or a combination of these and other factors offer plausible reasons to think that human use of fossil fuel reserves is likely to decline in most post-Anthropocene scenarios. Human control of the planet's surface and other animals is also likely to decline, whether because of shrinking human populations, breakdowns in organization required to control large amounts of surface area, or—as documented later—declines in energy availability that make motorized transportation over long distances increasingly difficult.

What would an energy-constrained world be like? What would humans still be able to do in this world and which capabilities would they lose? Future humans may be forced by their material circumstances to distinguish carefully between needs and wants and to develop a way of living that emphasizes sufficiency rather than luxury. Past human societies, which had less available energy than today's societies, provide some useful evidence about how to function in low-energy environments, and so it is possible to read accounts of earlier societies for guidance on this question.[2] Consulting past societies as models for future ones remains necessary but insufficient, however, for making observations about the possible paths of existing societies because (i) our understandings of past civilizations are imperfect, as the tendency of historians to revise these understandings every generation or so demonstrates, because (ii) post-Anthropocene humans will have some form of access to and knowledge about technologies and schemes for organizing society that pre-Anthropocene societies did not, and because (iii) it is unrealistic to expect present societies to simply revert to past forms, not least because things change as they move forward. History is not a time reversible process. Other work focused on the Anthropocene proper has discussed ways of conceptualizing an exit from the epoch and the material realities that occasion and condition such an exit.[3]

One way of making observations about future paths is to start from physical or material limits that create boundary conditions for what future societies will look like.[4] Taking a sufficiency-focused approach within these limits involves designing systems that are sufficient to support life without exceeding

1 Watson and Watson 2020, pp. 1–3.
2 Tainter et al. 2003; Hall et al. 2009, pp. 25–47.
3 Tainter 2011; Heikkurinen (ed.) 2017.
4 Heikkurinen et al. 2019.

certain material parameters. This paper represents one such attempt to work out these boundary conditions with respect to primary energy production dedicated to transportation, especially daily commuting. I estimate the energy, in gigajoules per person, that will be available for transportation under two energy-constrained scenarios. I then examine several modes of transportation (cars, metros, buses, bicycles, scooters, and walking) and calculate how much distance a person can cover with each given certain energy constraints. I discuss the implications of these estimates for the mode choice, speed, and physical size of human settlements in a world after human dominance.

The scenarios considered in this analysis focus on reductions in available energy, which may result from factors like climate change and fossil fuel depletion. However, there are other planetary boundaries, like freshwater availability, biodiversity, and land-use, that are not directly accounted for here.

2 Energy and Distance

Current primary energy consumption per person in the US is about 326 gigajoules (GJ). Worldwide, it is closer to 82 GJ per person.[5] This number captures the electricity consumed from power plants (including imported and domestic coal, nuclear, natural gas, hydroelectric, geothermal, solar [thermal, concentrated, and photovoltaic], wind, wood, biomass), from burning petroleum products like gasoline and diesel, and from burning ethanol and biodiesel, as well as 'non-combustion' fossil fuel use, as in construction, feedstock, and manufacturing.[6] Primary energy consumption, therefore, offers a reasonable measure of the energy required to sustain the lifestyle and activities of a given population. Its increase or decrease is a matter of general concern.

A future decline in the energy available to humans could follow a gentle slope: economies could transition to low-carbon energy sources like wind and solar quickly and thoroughly, major catastrophes could be averted, and humans could arrive at a steady state in which total available energy is lower but still high by historical standards. In IPAT terms, this would correspond to a declining technology multiplier that balances affluence and population holding steady (or increasing modestly). Or a decline in available energy could happen quickly. Humans' most current and most lucrative energy-production techniques (e.g., longwall mining, deepwater drilling, hydraulic fracturing) tend to rely on human dominance and complexity. This means that failures in

5 Steffen et al. 2018, pp. 8252–269; EIA 2020a.
6 EIA 2020a.

the systems these techniques rely upon could lead to sharp declines in available energy. In IPAT terms, this scenario would correspond to unpleasant decreases in population, affluence, and likely technology as well. It would be dire.

To quantify and explore the differences between these two possible futures, I model two scenarios that I call demi-Anthropocene and post-Anthropocene. The first, demi-Anthropocene (demi-A), represents a partial and optimistic exit from the human-dominated geological epoch. It is characterized by a holistic sufficiency approach to human consumption and energy use. It would involve, for example, the sort of energy transition described by Jacobson et al., who suggest that a reduction in total energy demand of around 57.1% is consistent with a transition to entirely carbon-neutral energy systems.[7] A decline of this magnitude would involve distinguishing human needs from human wants and essential energy use from frivolous energy use. Estimates of reductions in energy use vary among different analyses[8] but are generally within ten percentage points of each other and of Jacobson et al.'s estimate. I use the round figure of 45% of available energy per person in the United States (that is, a 55% reduction) in the demi-A scenario. It captures the median of such work and avoids clothing itself in a false precision. The demi-Anthropocene scenario represents the sort of world brought about by rapid shifts to renewable energy alongside the widespread electrification of transportation and industry. This happy demimonde would involve nation-states and companies aggressively installing renewable energy infrastructure in the next three decades, closing coal and gas power plants, and phasing out gasoline and diesel vehicles by 2060 or earlier. It would also involve humans consuming and demanding less as part of a sufficiency-minded approach. These changes would be rapid and radical but would not overturn the global economy or the total amount of energy available to human societies. Note, for instance, that because the demi-Anthropocene is based on 45% of the current US energy primary energy supply per capita, it would actually involve an *increase* in world energy supply per capita (from about 82 GJ to about 147 GJ per person per year). The demi-A scenario, therefore, can accommodate visions of increasing global standards of living for all but the most energy-intensive populations, who would have to reduce their energy demand by more than half, either through efficiency improvements or straightforward, sufficiency-minded reductions in goods, services, and travel. This scenario would also involve some cooperation from non-human factors: to land safely on a demi-A plateau, worst-case scenarios like runaway feedback loops from widespread Arctic methane release will

7 EIA 2020b.
8 Jacobson et al. 2017, 108–121.

probably have to be avoided, as they are likely to frustrate the supply chains, social systems, and technical improvements that will make providing 147 GJ of energy per person per year possible.

The second scenario, post-Anthropocene (post-A), is harder to schematize. This scenario represents a severe decline in available energy—for purposes of this analysis, I model it as a 95% drop from the total energy available in the US in 2018. The post-A describes a collapse of the human-dominated geological epoch and, after it, a world in which human control of the planet's energy has slipped away. In this scenario, societies could no longer harness the energy available on the planet as efficiently or at the scales they do today. They would be limited in their ability to organize and operate societies that demand high concentrations of energy. In climatological terms, such a scenario might coincide with, to take one example, the hotter Earth scenario presented by Steffen et al. (2018), in which humans live on in a world much warmer (say four to five degrees Celsius above preindustrial levels) and less hospitable to them. This scenario could also be caused by a global cooling event resulting from a nuclear exchange or other catastrophe. Unlike the demi-A scenario, this is a world in which global levels of available energy would shrink from 82 GJ to fewer than 17 GJ. Standards of living would fall for virtually everyone on the planet. Travel, the consumption of goods, and the provision of critical infrastructure services would become fragile and difficult. There are fewer ways of estimating the energy available in such a situation because these situations are more distant and more different from our own. No matter how many studies of the Eocene or a "hothouse Earth" are conducted, the quantity of available energy in such scenarios remains unknown. We do know, however, that energy will be much harder to come by. For purposes of this analysis, I set the post-A benchmark value at 5% of current available energy (that is, a 95% reduction in available energy from 2018 levels in the United States).

I mostly rely on US data in this analysis (noted when otherwise) because the United States represents the largest, most carbon-intensive economy whose figures on vehicle miles driven, total primary energy production, and other metrics tend to be comprehensive and accessible. I also start the analysis from US numbers because the United States has often served as a vanguard for the energy-intensive, anti-sufficient patterns of behavior (personal car use, high amounts of meat consumption, extensive suburban housing) and the infrastructure systems that realize those actions (industrial farming, extensive fossil fuel extraction and delivery, financial subsidies for larger homes, a globally-deployed military) that make the most difference in the energy-use figures of a society.

In the demi-A scenario, total primary energy per US resident would decline to about 147 GJ, down 55% from 326 GJ today. In the post-A scenario, each person would make do with 16 GJ, down 95% from today. Of course, not all this energy is spent on travel. From 2016 to 2018, transportation consumed 28.4% of total energy usage in the US, or about 95 GJ per person.[9] (This is up from 23% in 1965 and 26.5% in 1990. The absolute increase is larger still, as these are percentages of a growing total energy pie. In 1965, energy dedicated to transport was likely under 20 GJ per person.) This 28.4% transportation share leaves ~41.6 GJ to each person per year in a soft-transition demi-A scenario and under five GJ per person per year in a hard-transition post-A scenario. The question is how feasible each mode of transportation is under these energy constraints. A table (Table 9.1) summarizing the results of this work appears after the following discussion, which focuses on cars, intracity trains (metros), and human-powered modes like bicycles, scooters, and walking.

3 Cars

The average US resident drives 21,688 kilometers (13,476 miles) per year.[10] This represents an energy expenditure of about 77.5 GJ, using a conversion factor of 34.34 MJ per liter of gasoline and an average fuel efficiency of 10.4 liters per 100 kilometers.[11] This figure, divided by the ~92.6 GJ that is used per person for transportation in the US today, indicates that ~83.7% of the energy budget for transportation in the US goes toward car travel, which is in line with US commuters' current modal share (76.6% of commuters drive themselves to work and another 9% carpool).[12] The energy expenditure of a car that is driven 21,688 kilometers per year outstrips the demi-Anthropocene energy budget by 86% and the post-Anthropocene energy budget by ~1,585%, or about sixteen times over. Note that six GJ roughly represents the chemical energy of combusting one barrel (159 liters) of crude oil.

Even if cars dropped to zero percent of the US modal share, the energy spent on all remaining forms of transportation (about 15 GJ) far exceeds the travel energy budget in a post-A scenario. This implies that further reductions among modes of transit seen as low-carbon transit, like commuter rail and buses, may be necessary in extreme cases. It is possible that, under scenarios

9 US Bureau of Transportation (2020a).
10 FHWA (2020).
11 Prokopiak (2012).
12 US Bureau of Transportation (2015); US Bureau of Transportation (2020b).

in which the calories available to human beings in the form of food are limited, strategies like reducing walking in favor of biking (biking is somewhere between two to four times as efficient as walking, and can have around 30% lower carbon intensity per km, depending on diet),[13] reducing walking and biking together by, e.g., reducing the size of human settlements and increasing density, or some combination of these and other strategies will be necessary.

Under the demi-A scenario, distance driven would be energetically capped at about 32,000 km/year. This assumes an efficient internal combustion car, the most efficient of which can move one kilometer at a cost of about 1.3 MJ. Even this is likely impossible, as it requires directing every last joule of energy spent on travel into cars. A more realistic soft cap would land around 25,600 km/year. (This number assumes the current balance of energy shifts modestly toward non-car travel, with car travel commanding a reduced, but still high, 80% of the travel energy budget.) This would limit most single-occupancy commuters to a bit over 70 km of total driving per day, effectively limiting the distance between workplace, school, and home to under 35 km (~22 miles), or from White Plains, New York, to the southern tip of Manhattan, or a little less than the driving distance from Machakos, Kenya, to downtown Nairobi. This would put downward pressure on US car use but is unlikely to impose hard limits: the average round trip car commute is currently about 51 kilometers. Europe arrives under this limit, albeit not by much: drivers in the UK, for example, covered 34.8 km per day in 2013, consuming about half of what the demi-A scenario would allow.[14] (Chinese driving numbers are harder to come by, but data from Cox and Liu et al. indicate that kilometers driven in China remain ~42% lower than the US but are climbing rapidly.[15]) However, these figures remain within the quite optimistic demi-A scenario. The numbers become grave when considering a harder transition.

In a post-Anthropocene 95% reduction scenario, the hard cap for driving would be about 2,900 kilometers per person per year. Again, this comes from redirecting 100% of the 4.6 GJ energy budget for travel toward car travel and is therefore almost certainly an overestimate. The soft cap for driving would be, using the optimistic 1.3 MJ per person per kilometer energy intensity figure, about 1,035 kilometers. As before, this assumes a reduction in the energy budget directed toward driving down to 80%. (In absolute terms, this is still quite high.) This scenario limits drivers to just 7,900 *meters* of driving per day, limiting the distance between home and commuter destinations to 3,900 meters

13 Coley (2001).

14 BBC (2014).

15 Cox (2019); Liu (2017), pp. 12709–724.

or so. This is the length of one large airport runway. This, obviously, leaves little room for car commuting in a post-Anthropocene scenario. Even electric cars, which are more efficient than internal combustion engine (ICE) cars, are too constrained under a low-energy scenario to provide travel beyond a few kilometers. These distances are better covered by foot, bicycle, scooter, or another form of low-energy intensity travel for reasons of practicality if nothing else.

Battery-electric vehicles (BEVs) vary widely in their efficiency, as do internal combustion engine vehicles. A stringently efficient ICE car like the Volkswagen Polo expends about 1.3 MJ to move one kilometer. An efficient BEV, like the Nissan Leaf, expends about 0.67 MJ to move the same distance.[16] Using these efficiency numbers (which are still quite favorable to ICE cars), if 100% of the light-duty fleet were converted to electric vehicles, the post-Anthropocene soft cap for one-way feasible commute distance would rise to a still-modest 7,556 meters per person per day.

Keep in mind that these numbers do not incorporate the energy costs or material demands of manufacturing cars. A rough approximation of materials and energy expended for manufacture can be gleaned by looking at the weights of the relevant vehicles: most light-duty BEVs weigh between 1600 and 2500 kg, ICE cars between 1100 and 2500 kg, bicycles 7–16 kg, and shoes 0.2–1 kg. The manufacturing energy required roughly scales with the total weight of each vehicle. Cars take around 73 GJ to manufacture per unit.[17] A bicycle, based on weight alone, should take about half a GJ to manufacture. Note that, when amortized over a ~12 year lifespan, manufacturing energy is tiny for bicycles and modest for cars. Energy consumption from use is eight to ten times more important than energy consumption from manufacturing. Note, however, that if manufacture energy is included as part of the energy budget, one car, BEV or not, consumes five years of the post-Anthropocene energy budget via the energy demands of its manufacture alone. (Note further that this energy-only analysis does not touch on the availability of rare earth metals and other materials needed to produce the amounts of BEVs needed for mass electrification, material in little demand by bicycles and walking shoes.) Hydrogen cars fare worse than BEVs on energy intensity metrics, falling somewhere between BEVs and ICE vehicles.[18] In anything approaching a post-Anthropocene world, cars will require commutes so short that other uses of transportation are better suited even if one considers factors other than energy intensity.

16 EPA (2020).

17 Delucchi et al. (2020); Stodolsky (1995).

18 Łebkowski (2019); MacKay (2010).

If cars have little use where they are not impossible, how do other forms of transportation, like metros, fare?

4 Trains Inside and between Cities; a Note on Buses

I will examine mass rapid transit systems, or metros, next. Catling (1966) found that the London Underground expends 0.1584 MJ to move one passenger one kilometer. The physics of moving trains have changed little in the intervening years, and contemporary sources confirm his findings, generally offering values between 0.09 and 0.35 MJ per passenger-km. This is about one order of magnitude less than a very efficient ICE car and one quarter of the energy required by a BEV. It falls about halfway between walking, which is less efficient, and cycling, which is more efficient. Under the demi-Anthropocene scenario of 41.6 GJ available per person per year for transport, distance traveled via metro would be subject to a soft cap of around 210,000 km/year. (This number, as with the soft cap calculation for cars, assumes that rapid transit consumes 80% of the travel energy budget.) This is a hugely generous travel allowance, allowing each user to circle the globe several times per year, or to cover 575 km per day. For all practical purposes, rapid transit can accommodate arbitrarily long commutes. Travel distances would be limited by other constraints, like time, geography, and the energy and natural resource investment required to initially build out these systems and to maintain them. These numbers become more constrained in a post-Anthropocene world.

The post-Anthropocene scenario allows for 23,400 km per year of travel using mass rapid transit, or 64 km of total riding per day, with 20% of the travel budget left over for other modes (which, in a system dominated by metro use, will be necessary to cover "last-mile" scenarios and other gaps in the network. Note, too, that intercity rail transit is usually slightly more efficient than are metros, allowing for moderate travel between cities under this scenario.) This travel limit is no longer arbitrarily high, and this has real-world consequences: commuters will not have the energy budget to commute by rail between Washington, D.C., and Baltimore (about 125 km roundtrip via MARC, an intercity rail system that requires similar energy per kilometer as a metro) five days per week, for example, or from Potsdam to central Berlin (about 120 km roundtrip via S-Bahn), or between Shanghai and its outlying suburbs, or between Tokyo and its outlying suburbs.

As a comparison point, BEV and ICE buses are also included in the table below. They are less efficient than metros; Cushman-Roisin and Tanaka Cremonini (2019) find that standard diesel-powered urban buses use 0.4 MJ to move one passenger one kilometer if the bus carries 20 passengers, and 0.6 MJ if it carries ten. This is about 20% more efficient than a BEV and 58% more efficient than an ICE car.

Electric buses employ a battery-electric drivetrain with regenerative braking. They perform better than standard diesel buses. The efficiency advantages of battery-electric systems relative to combustion engines are compounded by the effectiveness of regenerative braking, which is able to recapture significant energy otherwise lost to brake heat in the start-and-stop use cases common to urban bus routes. Full electric buses remain more expensive than their diesel predecessors, but adoption is widespread in China, where about 90% of new bus sales are electric, and major cities like Shenzhen only use electric buses. Uptake has been slower outside of China, though a variety of major cities have committed to electric-only bus fleets by dates like 2035 (San Francisco), 2037 (London), and 2040 (New York City).

Electric urban transit buses consume between 0.18 and 0.4 MJ per passenger-km, depending on the analysis used.[19] I will use an average figure of 0.29 MJ per passenger-km. Under this assumption, distance traveled via electric bus would have a soft cap of around 115,000 km/year under a demi-A scenario. In an electric bus system, commuters would be able to cover 310 km per day without exceeding the energy budget for transportation. This distance covers virtually all commute scenarios. Travel distance, for electric buses, would be constrained by factors other than energy in a demi-A scenario.

A post-Anthropocene scenario allows for 12,700 km/year of travel using battery-electric buses, or 35 km of travel per day, with 20% of the travel budget remaining over for other modes (which, as before, will be necessary to cover last-mile scenarios and other gaps). This travel limit allows for most travel within cities, few of which are wider than 17.5 km, but remains somewhat restrictive for most other scenarios. Travel between Flushing, Queens, and lower Manhattan five days per week would exceed the energy available for transportation in such a model. Nor could anyone commute between, say, Cairo and New Cairo under these constraints. In fact, commuters would be hard pressed to commute from the suburbs of all but the most compact major cities.

Metros and electric buses, unlike cars, consume sufficiently low levels of energy in a post-A scenario to be feasible for daily commuting within human settlements no wider than about 30 km, or between settlements no further apart than 30 km. Their consumption of energy in a demi-A scenario allows for arbitrarily long commutes (over 285 km each way). However, this does not fully

19 Łebkowski (2019); Graurs et al. (2015).

account for the fact that metros are complicated and resource-intensive to build and maintain. In reality, the demands of repair and maintenance would make large metro systems more difficult to run in an energy-constrained scenario than their simple energy costs to run imply. In the next section, I turn to ways of getting around that require little infrastructure and maintenance.

5 Bicycles, Scooters, and Walking

Other than some electric kick scooters and solar-powered vehicles built for specific contests, the bicycle is the most energy efficient means of transportation known. This includes human-devised travel and locomotion among the land animals. A person on a bicycle, moving at 16 kilometers per hour, consumes between 0.095 and 0.115 MJ per kilometer per person.[20]

The soft cap for bicycle travel under a demi-A scenario is 317,000 kilometers. This figure doubles the world record for cycling in a year (which record was accomplished by biking twelve or more hours per day for a year). The bicycle's distance budget is inexhaustible: 25 kilometers per hour for 24 hours is only 600 km, still under the energetic limit. The distance limits for bicycle commuting are unconstrained by available energy under a demi-Anthropocene scenario. (In practice, the limit would be human willingness or, failing that, the human body's ability to absorb calories.) Much of the energy saved by bicycle commuting could eventually be reallocated to areas like manufacturing, medicine, or agriculture.

The soft cap for bicycle travel under a post-Anthropocene scenario, allocating 80% of energy to cycling, would be 35,200 km per person per year. This allows for 95 km of travel per person per day, or an effective commute limit of 47 km. Bicycle commuting, therefore, would remain impeded only by non-energetic reasons (not many people will commute 40 km on a bicycle, although an electric bike, safe infrastructure, and pleasant weather certainly make the prospect more appealing). Commutes between Texcoco and Mexico City, or Plano and Dallas, would remain possible seven days per week. Commutes between downtown Dallas and downtown Fort Worth would remain possible five days per week, as would travel between famously sprawl-friendly Houston and most of its outlying suburbs. In all but the most extreme cases, the bicycle keeps the suburbs within reach of the central business districts they feed. (The electric bicycle, which is only slightly less energy efficient than a regular bike, makes these commutes more practical and accessible.)

20 Pivit (1990); McArdle et al. (2000).

Ironically, given the historical intertwining of the car and the suburbs, the bicycle is the only mode of transportation that would make something like true sufficiency-minded suburbs possible in a post-Anthropocene world. This idea resonates with the arguments presented by Mikko Jalas in Chapter 3, which rethinks affluence in terms of time. In a low-energy future, the bicycle's ability to facilitate mobility while minimizing energy use could be a key factor in enabling slower, more time-affluent lifestyles and settlement patterns.

There is one final mode of transportation that is even more efficient than the bicycle. Curiously enough, it is the electric kick scooter.

Electric kick scooters are strikingly efficient. (To be clear, kick scooters are different from electric mopeds, which are more like small motorcycles.) An electric kick scooter consumes, on average, about 0.045 megajoules per passenger-km.[21] This is the lowest figure for any mode in this study. A demi-Anthropocene society in which 80% of the transportation energy budget is allocated to small electric scooters can supply enough energy for each of its members to move 740,000 km per person per year. This is seventeen times around the Earth. It is two thousand kilometers per day, or more than anyone can scoot in a week.

Under a post-A scenario, scooter users would be able to cover 82,000 km per year, or over two hundred km per day. Users could commute from Irvine to downtown Los Angeles, if for some reason they wanted to. Energy is unlikely to constrain scooters even under quite stringent conditions. But it is possible that under certain post-A scenarios scooters and bicycles could not be produced or maintained due to breakdowns in infrastructure or trade. This leaves walking.

Walking, because it does not take advantage of a wheel and ball bearing system to conserve momentum, is less efficient than riding a bicycle or scooter. Because it does not require any external weight or complication, it is more efficient than almost all forms of motorized transportation. Combining the global average adult weight (62.5 kg) with evidence from McArdle et al. (2000) on the energy demanded by walking yields an energy consumption rate of 0.218 MJ per pedestrian per km.

A demi-Anthropocene society in which 80% of the transportation energy budget is allocated to walking can supply the energy for each of its members to cover about 150,000 km on foot per person per year. This is 418 kilometers per day. As expected, energy is not a significant constraint for walking under a demi-A scenario. Practical limitations remain more important than considerations of energy expenditure.

21 Weiss et al. (2020).

TABLE 9.1 Energy intensity, distance, and predicted effective commute range (by mode)

	Energy intensity (MJ per passenger-km)	Distance/ year (km)	Distance/year (km)	Effective commute limit (km)	Effective commute limit (km)
		Demi-A	Post-A	One-way, demi-A	One-way, post-A
Cars (ICE)	1.3	25,610	2,843	35	3.9
Cars (BEV)	0.67	49,691	5,516	68	7.6
Metros	0.1584	210,182	23,353	288	32
Buses (ICE)	0.54	61,653	6,844	84	9.4
Buses (BEV)	0.29	114,803	12,745	157	17.5
Bicycles	0.105	317,074	35,200	434	48.2
Electric scooters	0.045	739,840	82,133	1,013	112.5
Walking	0.218	152,719	16,954	209	23.2

Under a post-A scenario, members of the same walking-dominant society would be able to cover 17,000 km per year, or about 45 km per day. This is at the limit of what a person can cover in one hard day of hiking. Walking is unlikely to be constrained by energy. Instead, it will be constrained by the usual things that hamper walking long distances: walking is slow (about 3–7 km/h), so it takes a long time to get anywhere further than a few thousand meters away, you are exposed to the elements while you do it, it is tiring, sensitive to terrain, and a burdensome means of transporting cargo.

6 Motorized and Non-motorized Transportation

Note that cars, metros, and buses draw from energy stores external to themselves, either via electricity or the pumping of liquid fuels like diesel and gasoline. Human-powered transportation, like biking and walking, draws on the metabolic processes of human beings and, therefore, from the food humans eat and stored energy in the form of fat. The unit we have been using so far, the megajoule, is equal to about 239 kilocalories. A cyclist who uses 0.1 MJ to travel one kilometer will require about 24 kcal of food, or about one walnut, to do so. This means that, unlike the motorized forms of transit, the diet of the person walking or cycling matters for the energy efficiency and environmental effects of these modes. A cyclist who mostly consumed beans, wheat, and corn, for example, could have an environmental impact as low as one-twentieth that of a beef-fed cyclist.[22]

Please note that the transportation budgets used in this paper are derived from primary energy production in the United States. These numbers carry more or less straight over to motor vehicles, which use primary energy sources like gasoline, diesel, and electricity. However, primary energy production does not directly include kilocalories consumed per person per day or some equivalent measure. The overall numbers for food energy are not too hard to calculate (e.g., US residents eat about 3300 kilocalories per person per day, which is about 5 GJ of food energy per person per year), but for consistency, this paper assumes that for societies in which large-scale conversions to non-motorized forms of transportation occur, the pool of primary energy currently used for transportation will, in some form, still be drawn on by non-motorized forms of transportation. This may involve, as a practical matter, transferring material and power from centralized electricity production or fossil fuel extraction toward increased calorie production, but the underlying energy budget will remain similar.

There are, finally, several reasons aside from energy intensity for preferring mechanically simpler and physically lighter forms of transportation. Cars, buses, and trains, whether combustion or battery-electric, require extensive infrastructure in the form of electrical grids, power plants, substations, mining, and machining. Bicycles and shoes require no fueling infrastructure beyond growing and transporting slightly more food. They also require lighter and fewer components for their production and maintenance. These economies can be estimated by looking at the curb weight of bicycles and shoes against the curb weight of motorized vehicles. Bicycles weigh between two or three orders of magnitude less than cars, buses, and metros do (even when divided by the average passenger load of each). Shoes weigh about four orders of magnitude less than those vehicles. We should expect, at a rough cut, a bicycle to demand about 1/500th the materials and complexity that a car, bus, or train would (on a per passenger basis). Similarly, we should expect to be able to produce thousands of pairs of shoes with the materials and manufacturing required for one car. In a post-Anthropocene scenario in which materials and supply chains are limited, these material economies would matter. Another way to put the point is: by forsaking luxury for some (the car, e.g.), future societies can achieve sufficiency for most (via bicycles and shoes).

Considerations of equity offer reasons to prefer simpler, lighter means of transportation. Because cars, buses, and trains are more expensive to produce

22 Clark et al. (2019); Roche (2020).

and use than bicycles and shoes (in ratios resembling their weight differences), they are more expensive for people to use. This is likely to translate preexisting inequalities into the transportation system, as is currently common. Fewer people can afford a car than can afford a bicycle. Whether the costs for complicated, resource-intensive transportation are paid upfront, financed via debt, or amortized via ticket sales, road taxes, and insurance, they fall heaviest on those least able to pay. The only exception to this fact would be a society that used some form of progressive taxation to subsidize motorized travel (either via public transit or, say, by subsidizing private car ownership).

7 Velocity and Physical Space

Changing a mode of travel's speed modifies the relationship this form of travel establishes between energy consumption and distance. At higher levels of speed, more energy is expended resisting the movement of air. As a result, running the same modes of transportation more slowly is one way of using a limited energy budget in a more sufficient manner. Hastening a mode of transit is, conversely, a way of trading energy for time. A cyclist traveling at 15 km/h requires a bit over 30 watts (or 30 joules per second), less than half of which is spent on air resistance.[23] A bicycle traveling at 30 km/h demands 148 watts. More than two-thirds of this power goes to moving air out of the way.

Pedestrians experience similar costs associated with speed, as do motor vehicles. A rough application of the drag equation indicates that a city bus moving at 30 km/h demands 1,875 watts (joules per second) to push through the air (assuming a typical drag coefficient of 0.6). The same bus moving at 60 km/h loses 15,000 watts to air resistance, or 15 kJ per second. Decreases in energy efficiency via higher speed reduce the effective range for all mode types under constrained-energy scenarios (with the exception of walking, which is still limited by factors like time and fatigue). If air resistance accounts for, say, 40% of the energy expended by an intercity bus running its route, and the energy spent overcoming air resistance increases eightfold when speed doubles from 30 km/h, something like a 35% reduction in range can be expected from running buses closer to 60 km/h as opposed to 30 km/h. Using range numbers from the post-A scenario implies that fast diesel buses would have a new effective commuting range below six kilometers each way and electric buses would be unable to cover commuting distances more than 11 kilometers each way. Because the energy required

23 Pivit (1990).

to overcome air resistance increases non-linearly, further speed increases shrink this range sharply. In an energy-constrained system, speed is incompatible with sufficiency.

Illich (1974) argues that the demands speed makes on energy tends to change societies' physical organization. These energy demands, and their effects, become particularly dramatic for means of transportation that go faster than ~24 km/h. Illich points out that certain levels of speed demand sufficiently high concentrations of energy as to require physical separation from other people. The space and infrastructure required changes the physical and social organization of a human settlements. He argues that 'the critical quantum' of speed is about 15 mph, or a bit over 24 km/h, and that '[w]herever this limit has been passed [a] basic pattern of social degradation by high energy quanta' can be observed. After this limit has been breached by public or private vehicles, 'equity decline[s] and the scarcity of both time and space increase[s]. Motorized transportation monopolize[s] traffic and block[s] self-powered transit.'

As we have seen, something important inheres in the relationship between speed and energy intensity, but a more detailed exploration of these claims is needed. This paper has attempted to analyze the specific energy intensity of different modes of transportation in a way that prices in their average velocities but also considers their various efficiencies and energy demands as really-existing systems of movement. This information allows for a reevaluation of Illich's observation. We do see some correlation between the modes of transportation with high energy intensities and high speed (Table 9.2).

TABLE 9.2 Energy intensity and typical effective speed

	Energy intensity (MJ per passenger km)	**Speed range usually achieved in cities**
Cars	0.67–5	30–95 km/h[24]
Metros	0.1584	25–85 km/h
Buses	0.29–0.54	13–85 km/h
Cycling	0.105	13–32 km/h
Scooters	0.045	10–32 km/h
Walking	0.218	3–7 km/h

24 Geotab (2023).

This correlation is somewhat jagged, however, with buses fairly inefficient for speed traveled, and metros and bicycles and scooters quite efficient. Cars remain much less efficient (even for their speed) than other ways of getting around. Illich's dictum captures an important insight about the relationship between speed, energy, and the physical and social organization of cities, but probably remains too simple. A bike or metro train at 30 km/h requires a different amount of power, and therefore a different organization of energy, than does a car traveling at that same speed. Both are over Illich's speed limit. But under the energy-constrained scenarios I have considered, a car cannot be sustained at that speed for the length of an average commute, while a train or bicycle can. Moreover, the amount of infrastructure and space that must be dedicated to cars traveling at 30 km/h through a city is much higher, on a per-passenger basis, than what must be dedicated to make it possible for subway cars and bicycles to travel 30 km/h through the same city. The reorganization of space to allow for increased energy-intensity travel has well-documented effects: neighborhoods delimbed by controlled-access highways, city centers covered by surface parking, pedestrians pushed back by road widening, playgrounds dusted with $PM_{2.5}$, and so forth. All of this is to say nothing of atmospheric carbon dioxide at 430 ppm and rising, ocean pH ~8.1 and falling, and September Arctic ice extent <5 million km^2 and shrinking.[25]

Under future conditions of energy constraint, Illich's posited relationship between energy intensity and the organization of cities may begin to work in reverse. Higher levels of speed will not drive increased extraction and the reordering of social and physical space. Instead, decreased extraction—either voluntarily as a result of a sufficiency approach or involuntarily as a result of material limits—will put downward pressure on speed. A decrease in aggregate speed will in turn change social space. As concrete examples, I have listed certain commuting routes that will be difficult or impossible for simple energy-expenditure reasons in a post-Anthropocene scenario. Because humans will be restricted to shorter, slower travel, human communities will experience pressure to become geographically smaller. This means that these communities will have fewer people, those people will be more densely packed, or some combination of the two. If humans are reluctant to give up the economic, social, and cultural benefits that come with living in organized settlements with many people,[26] they will have to choose density, probably with effective commuting distances much shorter than they are now.

25 Comiso et al. (2017); SCOR Biological Observatories Workshop (2009); Hall-Spencer et al. (2008); NASA (2023).

26 Tisdell (2008).

There is another set of scenarios, of course, in which humans give up on the agglomerative effects of living together and live in relatively isolated semi-rural communities with no complex system of transportation to speak of because no one has any commuting to do. I do not model this future, although I can say from the work here that, while bicycles are likely to be effective in these futures, it is hard to produce bicycles without some degree of agglomeration and trade. Bicycles require parts, especially in the drivetrain and bearings, that are machined with some degree of precision. Bikes generally require metallurgy, although bamboo- and wood-frame bicycles exist. Pneumatic tires require rubber and a way of compressing air. Most of these parts could be swapped out for simpler versions (e.g., solid tires instead of air-filled ones, a strider bike design instead of pedals and crank) but each swap would compromise the usefulness of the bicycle a bit further.

The analysis in this chapter suggests one motivating factor for policy change and several normative implications. The motivating factor comes from the fact that a demi-A scenario is probably not that bad and a post-A scenario is very bad. A lot of useful transportation technology (to say nothing of other socially useful goods) will be useless or severely limited in a world where the total energy available per human being is under 17 gigajoules. Human societies should make every effort to land on a higher energy plateau than on a lower one. This objective remains possible, but it will require rapid sufficiency-focused efforts to lower emissions and some degree of luck. The more rapid and effective the effort, the less luck will be needed. The first normative implication is that societies ought to take a sufficiency approach by planning for lower-energy states in ways that are far-reaching and detailed and that distinguish between human needs and human wants. Transportation is an obvious area, and one severely affected by energy constraints, but still only represents about 28% of total energy expenditure. Other areas, like manufacturing, healthcare, and communication technology, will be affected in much the same way as transportation, and will have to move toward a sufficiency approach that involves less total consumption paired with more efficient machines and system design. How many data centers can society maintain in a demi- or post-Anthropocene scenario? How many hospital ventilators per 100,000 people? Further research and planning in this area remain essential. A second recommendation is to model the intentional slowing of transportation systems and to build the possibility of slower, sufficiency-first, lower energy-expenditure operation into system design. A third recommendation is to take resources and physical space from high-energy systems of transportation to develop and encourage low-energy systems. A fourth recommendation is to plan for and construct denser human settlements. In addition to sparing energy and carbon emissions, density may be demanded by energy constraints. How would the place you live change if no one could commute more than ten kilometers in any direction? A final normative implication regards the necessity of planning for and investing in simpler, lighter systems to replace or supplement existing systems of transportation. These systems require less energy to manufacture, maintain, run, and repair, and are less likely to suffer breakdowns in the first place. They are therefore more sufficient in an energy-constrained world.

It is too much to expect that the sufficiency-focused downward pressure on transportation speed exerted by decreased energy availability will *reverse* the changes set into physical space by two centuries of fast, energy-intensive transportation. Future human settlements are unlikely to resemble a village idyll, because (i) history is not a time-reversible process and (ii) low-power technologies (e.g., information technology, ball bearings, pneumatic tires) that were limited or non-existent in the pre-velocity age will coexist with lower-energy forms of transportation in the sufficiency-first, post-velocity age. If the systems for knowledge transmission are more resistant to change and retrenchment than the systems for energy production, humans may pass into a state of relatively high knowledge paired with relatively low energy availability. Technologies that demand a moderate level of knowledge to design and build but a low level of energy to run, like the bicycle, are likely to play an increased role in such societies. The technologies that remain viable in a low-energy world can be expected to create new configurations of social and physical space. And new configurations of space loop back into the people who traverse that space, transforming attitudes and behaviors which further alter the built environment.[27] One can imagine, for example, suburbs centered on electric bicycle use, because no other mode can effectively reach city centers from outlying residential communities. Just as the car, combined with US urban planning, developed and spread specific forms of human settlement and patterns of energy across the planet in the twentieth century, the modes of transportation demanded by an energy-constrained system are likely to grow their own topologies of movement, exchange, and sufficiency.

8 Conclusion

It is impossible to know the knock-on effects of a human exit from planetary dominance. Most plausible scenarios involve a decline in human control of the material and surface area—and therefore the energy stocks and flows—of

27 Kidder (2009), pp. 307–28; Oldenziel and de la Bruhèze (2011), pp. 29–49.

Earth. This potential reduction in energy available to human societies will constrain the ability of these societies to move people through physical space, a task whose scale and complexity remain significant.

This article finds that the choice among different modes of transportation turns out to be non-trivial in energy-constrained scenarios. Energy-efficient means of transportation allow for physically larger human settlements, for example, within the limits of practicality. Bicycles and scooters allow for large coverage areas, rapid transit for moderately large ones, and cars for coverage areas so small they are better served by other means of transportation. Walking is not meaningfully energy constrained but is limited by considerations of time, terrain, and cargo. Transportation after the decline of current levels of energy availability is likely to be slower than existing transportation, even holding mode choice constant, because of the nonlinear relationship between velocity and the power needed to overcome drag. Finally, any post-Anthropocene scenario is likely to involve changes in the structures that preserve and communicate knowledge as well as those that produce and distribute energy. As a result, the transportation modes available to future societies will be different, the choices among them will be influenced by new considerations, and their development may rely on previously-developed knowledge sufficient to a low-energy world.

References

Catling, D.T. (1966) "Paper 8: Principles and Practice of Train Performance Applied to London Transport's Victoria Line," *Proc. Inst. Mech. Eng. Conf. Proc*. 181: 48–61.

Clark, M.A.; Springmann, M.; Hill, J.; Tilman, D. (2019) "Multiple health and environmental impacts of foods," *Proc. Natl. Acad. Sci. U.S.A.* 116: 23357–62.

Coley, D. (2001) *Emission Factors for Walking and Cycling*. Exeter, UK: Center for Energy and the Environment.

Comiso, J.C.; Meier, W.N.; Gersten, R. (2017) "Variability and trends in the Arctic Sea ice cover: Results from different techniques: trends in the arctic sea ice cover," *J. Geophys. Res. Oceans* 122: 6883–6900.

Cox, W. (2019) "Average Chinese Car Travels as Much as American Car," *New Geogr.*

Cushman-Roisin, B.; Tanaka Cremonini, B. (2019) Useful Numbers for Environmental Studies and Meaningful Comparisons. Unpublished.

Delucchi, M.; Burke, A.; Lipman, T.; Miller, M. (2020) Electric and Gasoline Vehicle Lifecycle Cost and Energy-Use Model. Available online: https://escholarship.org/uc/item/1np1h2zp.

Drivers' Annual Mileage Rates Fall. (2014) *BBC News*.

EIA. (2020a) How Much Energy Does a Person Use in a Year? Available online: https://www.eia.gov/tools/faqs/faq.php?id=85&t=1.

EIA. (2020b) Primary Energy Consumption. Available online: https://www.eia.gov/tools/glossary/index.php?id=Primary%20energy%20consumption.

EPA. (2020) Fuel Economy of the 2018 Nissan Leaf. Available online: https://www.fueleconomy.gov/feg/noframes/39860.shtml.

FHWA. (2020) Average Annual Miles per Driver by Age Group. Available online: https://www.fhwa.dot.gov/ohim/onh00/bar8.htm.

Geotab. (2020) Gridlocked Cities. Available online: https://www.geotab.com/gridlocked-cities/.

Graurs, I.; Laizans, A.; Rajeckis, P.; Rubenis, A. (2015) *Public Bus Energy Consumption Investigation for Transition to Electric Power and Semi-Dynamic Charging*. Jelgava, Latvia: Latvia University of Agriculture.

Hall, C.; Balogh, S.; Murphy, D. (2009) "What is the Minimum EROI that a Sustainable Society Must Have?" *Energies* 2: 25–47.

Hall-Spencer, J.M.; Rodolfo-Metalpa, R.; Martin, S.; Ransome, E.; Fine, M.; Turner, S.M.; Rowley, S.J.; Tedesco, D.; Buia, M.-C. (2008) "Volcanic carbon dioxide vents show ecosystem effects of ocean acidification," *Nature* 454: 96–99.

Hansen, K.; Breyer, C.; Lund, H. (2019) Status and perspectives on 100% renewable energy systems. *Energy* 175: 471–480.

Heikkurinen, P., Ed. (2017) *Sustainability and Peaceful Coexistence for the Anthropocene*. New York, NY: Routledge.

Heikkurinen, P.; Ruuska, T.; Wilén, K.; Ulvila, M. (2019) "The Anthropocene exit: Reconciling discursive tensions on the new geological epoch.," *Ecol. Econ.* 164.

Illich, I. (2009) *Energy and Equity*. New York, NY: Marion Boyars.

Jacobson, M.Z.; Delucchi, M.A.; Bauer, Z.A.F.; Goodman, S.C.; Chapman, W.E.; Cameron, M.A.; Bozonnat, C.; Chobadi, L.; Clonts, H.A.; Enevoldsen, P.; et al. (2017) "100% Clean and Renewable Wind, Water, and Sunlight All-Sector Energy Roadmaps for 139 Countries of the World," *Joule* 1: 108–121.

Kidder, J.L. (2009) "Appropriating the city: space, theory, and bike messengers," *Theor. Soc. 38*: 307–28.

Łebkowski, A. (2019) Studies of Energy Consumption by a City Bus Powered by a Hybrid Energy Storage System in Variable Road Conditions. *Energies* 12: 1–39.

Liu, H.; Man, H.; Cui, H.; Wang, Y.; Deng, F.; Wang, Y.; Yang, X.; Xiao, Q.; Zhang, Q.; Ding, Y.; et al. (2017) "An updated emission inventory of vehicular VOCs and IVOCs in China," *Atmos. Chem. Phys.* 17: 12709–24.

MacKay, D. (2010) *Sustainable Energy—without the Hot Air*. Cambridge, UK: UIT Cambridge.

McArdle, W.D.; Katch, F.I.; Katch, V.L. (2000) *Essentials of Exercise Physiology*, 2nd ed. Philadelphia, PA: Lippincott Williams & Wilkins.

NASA. (2020) Carbon Dioxide. Available online: https://climate.nasa.gov/vital-signs/carbon-dioxide/.

Oldenziel, R.; Albert de la Bruhèze, A. (2011) Contested Spaces. *Transfers* 1: 29–49.

Pivit, R. (1990) Measuring Aerodynamic Drag. *Radfahren* 2: 44–46.

Prokopiak, L. (2020) Cost Competitive Electric Vehicle Fleet. Available online: http://large.stanford.edu/courses/2012/ph240/prokopiak1/.

Roche, J. (2020) *Agribusiness: an International Perspective*. Philadelphia, PA: Routledge.

Rubin, T. (2020) Los Angeles Metro Bus System Compares Favorably With its Peer Group. Available online: https://www.newgeography.com/content/002361-los-angeles-metro-bus-system-compares-favorably-with-its-peer-group.

SCOR Biological Observatories Workshop. (2009) *Report of the Ocean Acidification and Oxygen Working Group*. Venice, Italy: SCOR.

Steffen, W.; Rockström, J.; Richardson, K.; Lenton, T.M.; Folke, C.; Liverman, D.; Summerhayes, C.P.; Barnosky, A.D.; Cornell, S.E.; Crucifix, M.; et al. (2018) Trajectories of the Earth System in the Anthropocene. *Proc. Natl. Acad. Sci. U.S.A.* 115, 8252–59.

Stodolsky, F.; Vyas, A.; Cuenca, R.; Gaines, L. (1995) *Life-Cycle Energy Savings Potential from Aluminum-Intensive Vehicles*. Cleveland, USA: International conference on beam processing of advanced materials.

Tainter, J.A. (2011) *The Collapse of Complex Societies*. Cambridge: Cambridge Univ. Press.

Tainter, J.A.; Allen, T.F.H.; Little, A.; Hoekstra, T.W. (2003) "Resource Transitions and Energy Gain: Contexts of Organization," *CE* 7 article 4.

Tisdell, C.A. (2008) *Economic Benefits and Drawbacks of Cities and Their Growth*. Brisbane, AU: University of Queensland.

US Bureau of Transportation Statistics. (2015) Commute Mode Share. Available online: https://www.bts.gov/content/commute-mode-share-2015.

US Bureau of Transportation Statistics. (2020a) Energy Intensity of Passenger Modes. Available online: https://www.bts.gov/content/energy-intensity-passenger-modes.

US Bureau of Transportation Statistics. (2020b) U.S. Energy Consumption by the Transportation Sector. Available online: https://www.bts.gov/content/us-energy-consumption-transportation-sector.

Watson, M.J.; Watson, D.M. (2020) "Post-Anthropocene Conservation," *Trends in Ecology & Evolution* 35: 1–3.

Weiss, M.; Cloos, K.C. & Helmers, E. (2020) "Energy efficiency trade-offs in small to large electric vehicles," *Environmental Sciences Europe* 32.

PART 4

Sufficiency in Population

CHAPTER 10

Taking Population Seriously in the IPAT Equation: Ethical and Political Implications of Planetary Boundaries

Michel Bourban

1 Introduction

The planetary boundary framework shows that the disruption of overarching, planetary-level systems are causing environmental changes that are taking us out of the stable environmental conditions of the Holocene. There are nine planetary boundaries—see Figure 10.1: the 'big three' (the climate system, the ozone layer, and the ocean); the four biosphere boundaries (biodiversity, land, fresh water, and nutrients); and the two aliens (novel entities and aerosols). Seven planetary systems have already been pushed beyond their critical limits: the climate system, biosphere integrity (biodiversity), the nitrogen and phosphorus cycles (nutrients), land use, freshwater and green water, ocean acidification, and novel entities (especially plastics).[1] If these transgressions persist, the entire planet may be pushed into a new state that would be much less hospitable for human societies, not to mention other species.

Climate change and biodiversity loss are especially crucial. On their own, they are pushing the Earth system out of the Holocene state: "transitions between time periods in Earth history have often been delineated by substantial shifts in climate, the biosphere, or both." Based on these observations, Earth system scientists suggest "a two-level hierarchy of boundaries, in which climate change and biosphere integrity should be recognized as core planetary

1 J. Rockström et al., "A Safe Operating Space for Humanity," *Nature* 461, no. 7263 (2009); W. Steffen et al., "Planetary Boundaries: Guiding Human Development on a Changing Planet," *Science* 347, no. 6223 (2015); L. Persson et al., "Outside the Safe Operating Space of the Planetary Boundary for Novel Entities," *Environmental Science & Technology* (2022); L. Wang-Erlandsson et al., "A Planetary Boundary for Green Water," *Nature Reviews Earth & Environment* 3, no. 6 (2022). Planetary Health Check 2025: https://www.stockholmresilience.org/news--events/general-news/2025-09-24-seven-of-nine-planetary-boundaries-now-breached.html, "Seven of nine planetary boundaries now breached".

DOI:10.1163/9789004746404_011

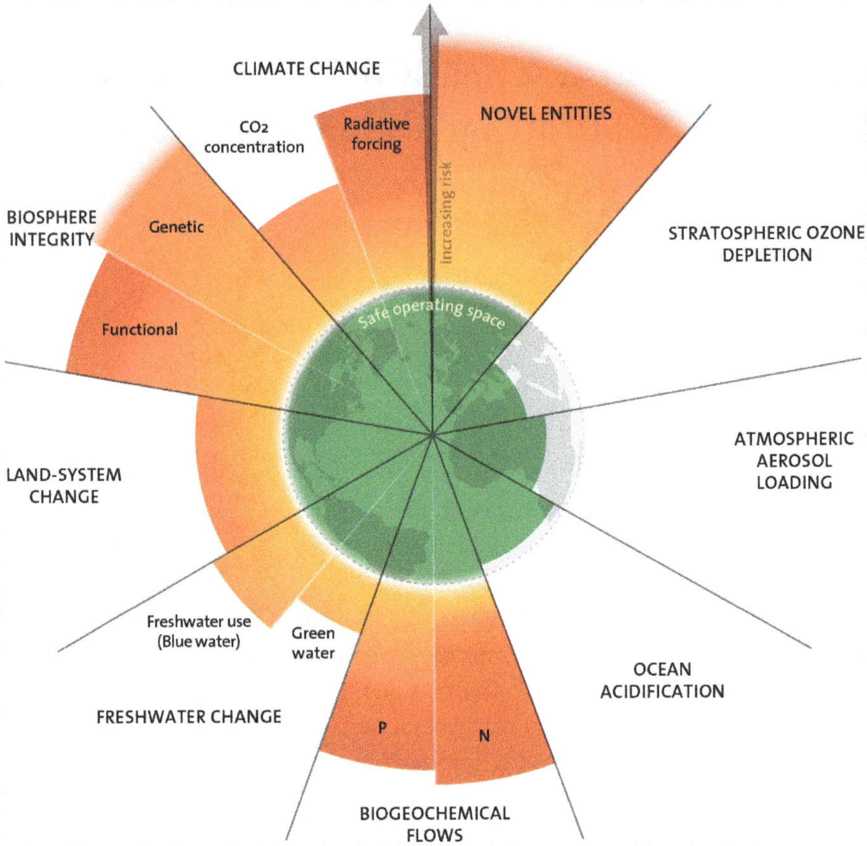

FIGURE 10.1 The nine planetary boundaries (2023 update). See also the 2025 update: https://www.stockholmresilience.org/news--events/general-news/2025-09-24-seven-of-nine-planetary-boundaries-now-breached.html, "Seven of nine planetary boundaries now breached" (Planetary Health Check 2025).

boundaries through which the other boundaries operate".[2] Crossing the planetary threshold of 2°C could lead the entire Earth system into a "Hothouse Earth" trajectory, in which global warming may be substantially accelerated.[3]

The Hothouse Earth pathway would be triggered by a cascade of tipping points in the climate system. Tipping points are self-perpetuating changes

2 Steffen et al., "Planetary Boundaries," 744.
3 Steffen et al., "Trajectories of the Earth System in the Anthropocene," *Proceedings of the National Academy of Sciences* 115, no. 33 (2018).

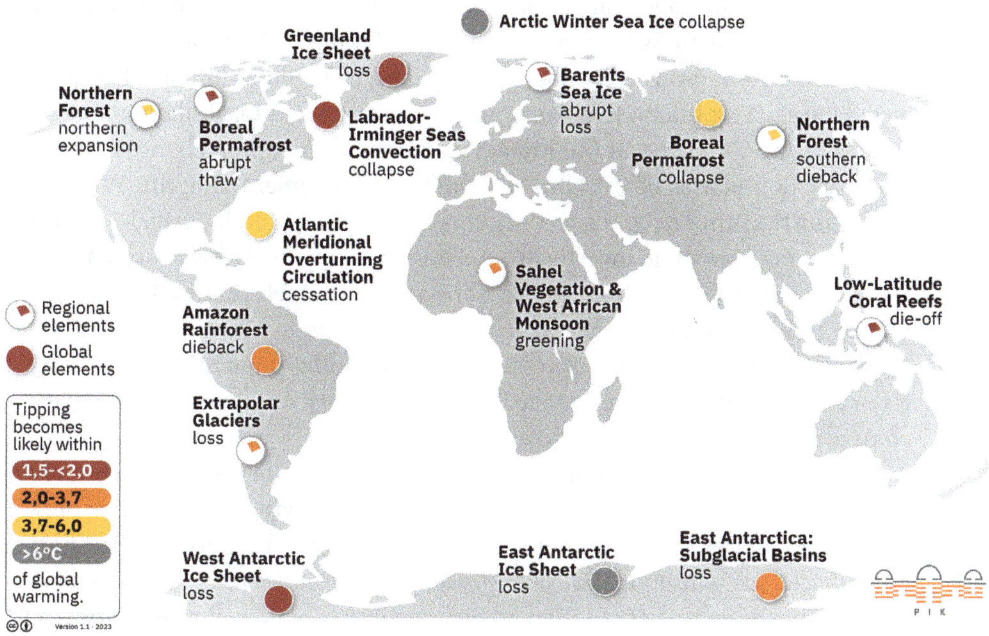

FIGURE 10.2 The geographical distribution of global and regional tipping elements, color-coded according to the best estimate for their temperature thresholds, beyond which the element would likely be 'tipped'
FIGURE DESIGNED AT PIK (UNDER CC BY LICENSE), BASED ON ARMSTRONG MCKAY ET AL., (2022)

beyond a warming threshold that lead to substantial and widespread impacts in the Earth system. The nine core tipping points are the two ice sheets in Greenland and West Antarctica, the Arctic and East Antarctic sea ice, the Boreal forest and the Amazon rainforest, Atlantic circulation, permafrost in Siberia, and Australia's Great Barrier Reef.[4] Each tipping point has a respective warming threshold (see Figure 10.2), but Armstrong McKay et al. (2022) have recently found that current global warming (about 1.1°C above pre-industrial levels) lies withing the lower end of five climate tipping points. This means that we already are at risk of triggering tipping points. Six tipping points become likely within the Paris Agreement range (1.5°C to <2°C), including the collapse of the Greenland and West Antarctic ice sheets, the die-off of low latitude coral reefs, and widespread and abrupt permafrost thaw.

Among the factors that contribute to the crossing of planetary boundaries, population growth plays a key role. World population has grown dramatically in recent decades, contributing substantially to the Great Acceleration: the

4 T. Lenton et al., "Climate Tipping Points—Too Risky to Bet Against," *Nature* 575 (2019).

major changes that have taken place in the socio-economic and biophysical spheres of the Earth System since the 1950s. From 2 billion people in 1930, world population grew to 4 billion in 1970, and to 8 billion in 2022. It is projected to reach 8.5 billion in 2030, 9.7 billion in 2050, and 10.4 billion in 2100.[5]

This chapter focuses on the impact of these demographic changes on the planetary boundaries of the climate system and biosphere integrity. Without neglecting the other major causes of climate change and biodiversity loss, such as economic growth, it investigates an essential but relatively marginalized aspect of the problem. It develops a normative approach that evaluates population policies and examines which policies can be justified in our context of planetary boundaries transgression, understanding population policies as government policies that influence fertility rates, such as educational policies, contraceptive availability, and taxes.

While green political theorists have been stressing for decades that truly sustainable societies require limits to population growth, this topic has only recently emerged in the philosophical literature.[6] Whether they support rights-based or consequentialist approaches, more and more philosophers have begun addressing population growth and its impacts on the environment, especially with regards to climate change.[7] As we will see below, while several climate ethics scholars such as Cafaro and Rieder are now indeed supporting population policies that reduce fertility, influential voices such as Caney and Heyward still tend to marginalize this mitigation measure in favor of other measures related to affluence levels and technological innovation.

A major reason why population growth reduction has only started to emerge in climate ethics scholarship over the last decade and remains marginalized is that this topic is unpopular. It raises many ethical and political problems and meets with strong social resistance.[8] But the problems specific to our context of population growth in a world of more than 8 billion human beings cannot be ignored. The point here is to foster democratic debate on this controversial

5 UN, "World Population Prospects 2022: Summary of Results," (New York: UN DESA, 2022).

6 It is true that the field of population ethics, which focuses on the number, identity, and quality of life of future people (and animals), emerged in the 1980s (see especially Parfit [1986]); the impact of population growth on the environment however represents a much more recent subject of concern for philosophers.

7 In this perspective, it is interesting to compare Cafaro's early review of the literature in 2012, which highlighted that "population policy has attracted relatively little attention from ethicists", with his more recent review, which stresses that "philosophical ethicists have, over the past decade, taken up the issue of climate change and population policy and there are now numerous articles and books which explore the subject." See Cafaro (2012, 45); Cafaro (2022, 1).

8 The way population growth has been overlooked in the last Summary for Policymakers of the Working Group III of the IPCC is a good illustration of this resistance: see Cafaro, chapter 12, this volume.

but no longer avoidable issue. In other words, this paper promotes public reasoning on complex political and normative issues that cannot be neglected, if we hope to keep climate disruption and biodiversity loss from crossing a critical threshold that could lead to the creation of a dangerous world.

The first section sets the framework for the analysis by explaining the contribution of the demographic factor to climate change and biodiversity loss. The second section critically examines an influential way of discussing population growth in the climate ethics literature: using the IPAT equation (Environmental impact = Population × Affluence × Technology) to over-emphasize the importance of "T" and downplay the role of "P." The third section explores and evaluates population policies and proposes possible justifications for certain positive and negative incentive measures. The final section considers objections commonly made against these incentive policies.

The core claim is that all three components of the IPAT equation need to be addressed urgently and simultaneously if we are to take seriously the radical implications of the planetary boundary framework. Technological innovation and affluence are indeed key to reduce the environmental impact of our lifestyles and policies and reach sustainable societies, but they cannot substitute for a reduction in the world's population.

2 Population Growth, Climate Change, and Biodiversity Loss

2.1 *The Demographic Factor*

Over the last decades, global GHG emissions have grown at a rate unprecedented in human history. Between 1970 and 2000, the average atmospheric concentration of GHGs increased by 1.3% per year. Between 2000 and 2010, it increased by 2.2% per year.[9] Average annual GHG emissions between 2010 and 2019 were higher than in any previous decade, even though the rate of growth slowed to 1.3% per year.[10]

Already in the AR5, the IPCC (2014, 8) stressed that population growth and climate change are directly linked: "Globally, economic and population growth continue to be the most important drivers of increases in CO_2 emissions from fossil fuel combustion." Indeed: "Between 2000 and 2010, both drivers outpaced

9 IPCC, "Summary for Policymakers," in *Climate Change 2014: Mitigation of Climate Change. Contribution of Working Group Iii to the Fifth Assessment Report of the Intergovernmental Panel on Climate Change*, ed. O. Edenhofer, R. et al. (Cambridge University Press, 2014), 6.

10 IPCC, "Summary for Policymakers," in *Climate Change 2022: Mitigation of Climate Change. Contribution of Working Group Iii to the Sixth Assessment Report of the Intergovernmental Panel on Climate Change*, ed. P. R. Shukla, et al. (Cambridge University Press, 2022), 10.

emission reductions from improvements in energy intensity." In the AR6, the IPCC also highlights that "Globally, GDP per capita and population growth remained the strongest drivers of CO_2 emissions from fossil fuel combustion in the last decade (*robust evidence, high agreement*)." Again, it stresses that "This growth outpaced the reduction in the use of energy per unit of GDP (–2% yr-1, globally) as well as improvements in the carbon intensity of energy (–0.3% yr–1)".[11] To reduce global GHG emissions, at least one of the two sources of growth must therefore be reduced, and probably both, since they are strongly interrelated.[12]

Population growth also plays a major role in biodiversity loss. According to the IPBES (2019, 5), there are five "direct drivers" of biodiversity loss: changes in land and sea use, overexploitation of species, climate change, pollution, and invasive species. Population growth is listed as an "indirect driver" of change, together with consumption patterns, trade, and technological innovation. Cafaro, Hansson, and Götmark (2022) reviewed over 30 recent studies that report a negative impact of excessive population on biodiversity. Population increases led to huge agricultural conversions of land for human use and phenomena such as urbanization and sprawl to accommodate human needs, which led to habitat loss; they also contributed to the overexploitation of species through overhunting, overfishing, and overharvesting; finally, they also caused the spread of invasive species, increased water and air pollution, and, as we saw above, contributed to increasing GHG atmospheric concentrations.

When discussing the importance of the demographic factor, it is important to avoid two pitfalls: (1) focusing only on population growth, isolating this factor from other drivers of environmental degradation; and (2) focusing only on developed or developing countries, leaving one of the two categories out of the analysis.

Regarding pitfall (1), population growth is not the only—and not necessarily the most important—driver of climate change and biodiversity loss. One of the key messages of the IPAT equation and its variations, such as the Kaya identity (Emissions = Population × GDP per capita × Energy used per unit of GDP × CO_2 generated per unit of energy), is that a multicausal analysis is required to properly understand current environmental problems.

Take climate change. Although the number of people matters, other factors, such as income, wealth, diets, education, household size and geographical

11 S. Dhakal et al., "Emissions Trends and Drivers," in IPCC (2022), 4.

12 This is why policies to slow down economic growth should also be implemented to reduce the deterioration of the biosphere's living conditions: see in particular the circular society model developed by Arnsperger and Bourg (2017) and the new environmental economics model supported by Laurent (2020).

location all lead to substantial variations in individual GHG emissions.[13] For instance, there is a positive relation between one's level of affluence and one's carbon footprint. According to Oxfam (2015), at the global level, the richest 10% are responsible for about 50% of GHG emissions, while the poorest 50% are responsible for only 10% of these emissions. The average carbon footprint of the richest 10% of the world's population is 11 times higher than the footprint of the poorest 50%, and 60 times higher than that of the poorest 10%. The average carbon footprint of the richest 1% of the world's population is about 175 times greater than that of the poorest 10%. Although a growing share of the richest 10% now live in emerging countries such as China, the overwhelming majority still live in OECD countries, with about one third in the US. For example, the per capita emissions of the richest 10% in China remain considerably lower than those of the richest 10% in OECD countries, partly because a large proportion of Chinese emissions come from the production of goods consumed in rich countries.

This radically unequal allocation of per capita emissions at the global level implies that the reproductive choices of people in developed countries have a much greater environmental impact than the reproductive choices of people in developing countries (at least regarding carbon emissions and climate change).[14] For instance, the total carbon footprint of a child born in the US is more than *160 times* greater than the footprint of a child born in Bangladesh; it is *200 times* greater than that of a child born in Niger, a country with a fertility rate of 7.6, the highest in the world.[15]

This leads us to pitfall (2). Given the highest carbon footprint associated with reproductive choices in developed countries, it might be tempting to focus on population policies exclusively in these countries. This would also avoid the problems associated with Malthus- and Hardin-inspired positions that place the overwhelming burden on poor countries in the name of environmental

13 J. Cohen, "Population and Climate Change," *Proceedings of the American Philosophical Society* 154, no. 2 (2010); L. Jiang and K. Hardee, "How Do Recent Population Trends Matter to Climate Change?," *Population Research and Policy Review* 30, no. 2 (2011).

14 It is important to stress that there are also important differences in per capita emissions within developed countries: not all citizens in developed countries are high GHG emitters, and there are more and more high GHG emitters in developing and emerging countries. See Ritchie and Roser (2019), OurWorldInData (2022).

15 P. Murtaugh and M. Schlax, "Reproduction and the Carbon Legacies of Individuals," *Global Environmental Change* 19, no. 1 (2009), 18; T. Rieder, *Toward a Small Family Ethic: How Overpopulation and Climate Change Are Affecting the Morality of Procreation* (Springer, 2016), 31.

protection.[16] Such a restrictive approach would, however, have two shortcomings. First, most population growth in coming decades is projected to occur among poor people in developing countries.[17] Demographic growth in developing countries also contributes to climate change, even if the birth of a child in a developing country has a smaller carbon footprint than the birth of a child in a developed country. Since GHG emissions need to be reduced everywhere to avoid a cascade of tipping points that would lead to a Hothouse Earth pathway, population growth should be limited everywhere. Second, as explained above, high fertility rates also contribute to biodiversity loss. Smaller human populations are necessary to preserve and restore biodiversity both in developed and developing countries. To sum up, as the total population in developing countries grows, the per capita environmental impact of people living in these countries is also very likely to grow. As a result, population growth must also be slowed in these countries to reduce the risks of crossing tipping points in the Earth system. For this reason, this chapter supports population policies both in developed and developing countries.

A legitimate concern raised by the inclusion of the environmental impact of population growth in developing countries is that it can lead to unjust outcomes: one should also consider other crucial factors, for instance in terms of social justice, such as the fact that having large families might be a necessity for poor people to protect their fundamental interests. But if the objective is also to address social injustices, reducing population growth everywhere remains a relevant measure. Not only does high fertility perpetuate poverty by locking more people into poverty traps; it also leads to environmental degradation, which harms poor people disproportionately.[18] Higher population paths could expose hundreds of million more people to climate impacts such as flooding, water stress, and drought over this century.[19] These numbers would be much higher if we cross tipping points in the Earth system. For this reason, a applying sufficiency to population policies is a way of protecting basic needs, interests, or rights, both now and in the future.

16 For a critical discussion of this type of approach, see Gardiner (2011, 443–456).

17 UN (2022).

18 J. O'Sullivan, "Synergy between Population Policy, Climate Adaptation and Mitigation," in *Pathways to a Sustainable Economy: Bridging the Gap between Paris Climate Change Commitments and Net Zero Emissions*, ed. M. Hossain et al. (Springer International Publishing, 2018).

19 J. Dodson et al., "Population Growth and Climate Change: Addressing the Overlooked Threat Multiplier," *Science of The Total Environment* 748 (2020).

2.2 *The Individual Level*

How could population policies contribute to mitigating climate change and restoring biodiversity? Let us start with the individual level.[20] Although most individual actions have an infinitesimal effect on the amount of GHGs in the atmosphere, reproductive choices have a substantial effect on individual GHG emissions by creating a "carbon legacy." Murtaugh and Schlax (2009) have quantified the impact of Americans' reproductive choices on individual emissions. According to their calculations, each child an average American chooses to have will most likely offset all of the emissions reductions they have previously made, even if they make significant efforts to limit their carbon footprint. In one likely scenario, each child adds on average 9,441 metric tons of CO_2 to an American's individual emissions, a carbon legacy that represents almost six times the parent's lifetime emissions.

Choosing to have one fewer child would save individuals in developed countries an average of 58.6 tons of CO_2-equivalent (tCO_{2e}) per year, which is considerable.[21] Indeed, this action has by far the greatest effect on reducing someone's individual carbon footprint. Other high-impact actions, such as living without a car, avoiding a transatlantic flight, and adopting a vegan diet, save respectively 2.4 tCO_{2e}, 1.6 tCO_{2e}, and 0.8 tCO_{2e} per year. Actions commonly recommended by governments and public institutions, such as using a hybrid car, recycling, and using energy-efficient light bulbs, are all comparatively low-impact actions, saving less than 0.8 tCO_{2e} annually.[22]

These empirical studies are based on a controversial notion: that of the carbon legacy. According to Murtaugh and Schlax, "a mother and father are each responsible for one half of the emissions of their offspring, and 1/4 of the emissions of their grandchildren".[23] A first difficulty with this notion is to find a fair

20 This sub-section assumes that individual agents are to some degree responsible to mitigate their own emissions. It is now well recognized in the climate ethics literature that individuals have duties to at least take steps to mitigate their emissions, but climate ethics scholars still disagree on why they have such duties and the stringency of these duties (Fragnière 2016). The idea here is not to support that individuals are unconditionally responsible for their emissions, but that they have a responsibility to reduce their carbon footprint as long as this can be reasonably expected of them, based on factors such as their financial capacity and the availability and affordability of energy and transportation infrastructures; see Bourban (2022, 2023).

21 S. Wynes and K. Nicholas, "The Climate Mitigation Gap: Education and Government Recommendations Miss the Most Effective Individual Actions," *Environmental Research Letters* 12, no. 7 (2017). Note that Wynes and Nicholas rely here on Murtaugh and Schlax (2009).

22 Wynes and Nicholas, "The Climate Mitigation Gap."

23 Murtaugh and Schlax, "Reproduction," 14.

allocation of responsibility between the parents and the children. Instead of attributing the responsibility for all of their children's emissions to the parents, an alternative would be to allocate the children's emissions to their parents as long as the former are not in a position to make their own choices, but to attribute the largest share of the responsibility for their own emissions to the children once they become independent. It is possible to debate at what point children really become independent from their parents, and it might be necessary to do so on a case-by-case basis, but in most cases, children might be considered old enough to start making their own choices at between 18 and 20 years old and should therefore assume most of the carbon footprint that comes with those choices. This could also avoid the risk of pushing children and grandchildren to think that they are off the moral hook, since their parents and grandparents have to assume the responsibility for the totality of their carbon footprint.

Another advantage of this alternative interpretation of the notion of carbon legacy is that it avoids the issue of the overdetermination of responsibility. Let us take three overlapping generations: generation A, B, and C. If we follow Murtaugh and Schlax's interpretation, the responsibility for the grandchildren's emissions (generation C) is overdetermined: their parents (generation B) are together responsible for 100% of their emissions, and their grandparents (generation A) for 50% of their emissions, leading to a total intergenerational responsibility of 150%. There is no such implication with the alternative interpretation.[24]

A second difficulty with the notion of carbon legacy is the uncertainty concerning the per capita emissions of children and grandchildren in the coming decades. Murtaugh and Schlax compare three possible scenarios relative to the 2005 baseline year, when global emissions were 4.31 metric tons of CO_2 (tCO_2) per person per year.[25] The optimistic scenario is based on aggressive GHG mitigation measures that would reduce per capita annual emissions to 0.5 tCO_2 by 2100; the constant scenario is based on maintaining emissions at 4.31 tCO_2 per capita per year throughout the century; the pessimistic scenario assumes that no drastic measures to combat climate change will be taken over the century,

24 Overdetermination of responsibility may be justified in some instances, but even if it were the case here, it would add a burden of justification to an argument that is already controversial enough: individuals' reproductive choices come with a carbon legacy they have a responsibility to assume. If there is an alternative way of looking at carbon legacy that does not involve overdetermination of responsibility, then this alternative should be pursued, if only for pragmatic reasons.

25 Murtaugh and Schlax, "Reproduction," 16.

resulting in a 50% increase in emissions per capita and per year in 2100 compared to 2005 (6.46 tCO_2).

Murtaugh and Schlax base their calculation on the second scenario, which represents an intermediary between "optimistic" and "pessimistic" projections. They note that this scenario does not take into account the likely increase in global GHG emissions in the first part of the century, implying that this brings it closer to an optimistic scenario. However, this scenario also does not take into account the possibility of a gradual reduction in per capita emissions over the next decades caused by an energy transition at the global level.

Despite these difficulties with the normative and empirical assumptions on which the notion of carbon legacy relies, Murtaugh and Schlax's approach has a double advantage. First, it gives a general idea of the impact that the decision to have a child can have in terms of additional GHG emissions in developed countries. Second, it shows that the emission reductions made possible by having one fewer child are substantial compared to most, if not all other actions. This remains the case even if one reduces the degree of responsibility of parents for their children's and grandchildren's emissions, and even if one chooses a more optimistic scenario for future GHG atmospheric concentrations.

2.3 *The Global Level*

Moving from the individual to the global level, O'Neill et al. (2010) have studied the effect of reduced population growth on future emissions, compared to the population growth projected as most likely by the dominant models. Under different scenarios of reduced population growth over the century, global GHG emissions are reduced by 16–29% by 2050, and by 37–41% by 2100. They conclude that reducing population growth would contribute significantly to the international effort not to exceed the 2°C limit for total average warming.[26]

Complementing the "wedge approach" by Pacala and Socolow (2004), where each wedge in a "stabilization triangle" represents 1 billion metric tons of carbon emissions avoided, Cafaro (2011, 208) calculated that holding global population to 8 billion in 2060 would provide about 3 wedges, meaning that it would prevent 3 billion metric tons of carbon from being released into to atmosphere. Reducing population growth would reduce more GHG emissions than the following five measures *combined*: doubling the fuel efficiency of the world auto fleet; halving the average mileage traveled per car; tripling the

26 These findings are confirmed by Rogelj et al. (2018, 327) and van Vuuren et al. (2018, 392), which stress the need to reduce population growth across the globe to at least 6.9 billion by 2100 if dangerous climate change is to be avoided.

number of nuclear reactors around the world; increasing current wind energy capacity 50 times; and halting global deforestation.[27]

Population decreases can also contribute to preserving other species or wild places and restoring biodiversity, in both developed and developing countries. Thanks to Europe's population stabilization, within the past two decades, about 7.6 million hectares of agricultural land have gone out of production in Eastern Europe, which allowed wild species to recolonize many former agricultural areas. Ecological restoration can accelerate and lock in these trends.[28] Over the past 50 years, successful national family planning programs in the Far East, South Asia, Latin America, Africa, and elsewhere have contributed to the conservation of biodiversity.[29] Four major empirical studies calculated that a sustainable global population would probably not exceed 3 billion individuals.[30] Even then, to be sustainable, societies would need to drastically improve existing modes of consumption and production. In addition, this number only considers Earth's human carrying capacity; leaving enough habitat and resources to avoid a mass species extinction would drive the number for a sustainable global population even lower.

3 How to Interpret the IPAT Equation?

How have climate ethicists reacted to these more and more numerous empirical studies on the environmental impacts of population growth? I discuss here

27 Cafaro (2012, 47). Cafaro draws here on O'Neill et al. (2010). Of course, this does not mean that these other measures should not be implemented as well—with the exception of developing new nuclear power plants: see Bourban (2022). Since reducing population growth alone will not be sufficient to maintain global temperature increases well below 2°C, these additional actions are also important. However, they do not have an emission reduction effect comparable to reducing population growth, unless they are all implemented together. As the next section explains, all these measures should be implemented quickly and jointly.

28 Cafaro, Hansson, and Götmark "Overpopulation."

29 W. Robinson and J. Ross, *The Global Family Planning Revolution* (The World Bank, 2007); R. Engelman and D. Johnson, "Removing Barriers to Family Planning, Empowering Sustainable Environmental Conservation: A Background Paper and Call for Action," in *Margaret Pyke Trust* (2019).

30 T. Lianos and A. Pseiridis, "Sustainable Welfare and Optimum Population Size," *Environment, Development and Sustainability* 18, no. 6 (2016); C. Tucker, "A Planet of 3 Billion," *Atlas Observatory* (2019); P. Dasgupta, *Time and the Generations: Population Ethics for a Diminishing Planet* (Columbia University Press, 2019); L. Tamburino and G. Bravo, "Reconciling a Positive Ecological Balance with Human Development: A Quantitative Assessment," *Ecological Indicators* 129 (2021).

two similar rights-based approaches that are representative of philosophical discussions on the topic.

According to Heyward (2012) and Caney (2020), population is certainly an important component of the IPAT equation, but they hold that we should focus first and foremost on the two other causes of disruption to the climate system: affluence and especially technology.

Heyward emphasizes the role of technological innovation in mitigation. Although dealing with population increase is, in her view, desirable, she focuses primarily on measures in developing countries and argues that the most important is to ensure compliance with the "collective duty to invest in technologies that could help accommodate a growing population".[31] Her main concern is that population reduction policies will lead to internalization, which holds people liable for the full costs of their reproductive choices. To avoid the risk of parents being forced to provide for their children's welfare on their own, she favors technological solutions, such as increases in energy efficiency and in the use of renewable energy. She adds, "Another potentially beneficial form of technology is carbon dioxide removal (CDR)".[32]

Caney also emphasizes technological factors such as increased energy efficiency, technology transfer, and the development of renewables and electric vehicles. He endorses the view of ecological liberalism, according to which "Persons should limit their environmental impacts in order to fulfill their responsibilities to future generations but can choose how to discharge this responsibility (selecting whether to do it through lowering consumption, enhanced use of clean technology, or limiting reproduction)".[33] This allows each person to fulfill their individual duty to reduce their carbon footprint as they see fit, in accordance with their life plan.

These positions have two main advantages: they avoid dealing with the controversial topic of population policy, which reduces their burden of justification in terms of practical solutions to climate change; and they appear to square reasonably well with basic moral intuitions in terms of individual rights and liberties. They however have two important limitations.

The first limitation is highlighted by the implications of the data discussed in the previous section. While the most recent empirical studies do mention

31 C. Heyward, "A Growing Problem? Dealing with Population Increases in Climate Justice," *Ethical Perspectives* 19, no. 4 (2012), 707.

32 Heyward, "A Growing Problem?", 722.

33 S. Caney, "Human Rights, Population, and Climate Change," in *Human Rights and 21st Century Challenges: Poverty, Conflict, and the Environment*, ed. D. Akande, et al. (Oxford University Press, 2020), 359.

technological innovation and consumption changes in addition to reduced population growth, they show that all these measures must be implemented *jointly* and *quickly* if we want to keep global temperature increases well below 2°C. The point is not to freely choose between one or the other of these measures, as ecological liberalism advocates, but to deploy them *simultaneously*. Without social, economic, and political institutions that create the required conditions for all these solutions to be implemented together, the 2°C planetary limit is likely to be crossed, and mutually reinforcing tipping points in the climate system are likely to be triggered, which could set the whole planet on a Hothouse Earth trajectory.[34]

This brings us to the core claim of this chapter: the IPAT equation should not be used to emphasize any one of its three factors over the others; it should rather be used to fully consider each of them. Without an institutional framework to promote technological innovation, consumption reductions, and population reductions through education, regulation and incentives, we risk creating a far more dangerous world. Earth system scientists point out that such a world is not necessarily a scenario for the distant future: "the evidence from tipping points alone suggests that we are in a state of planetary emergency: both the risk and urgency of the situation are acute." Indeed, "the intervention time left to prevent tipping could already have shrunk towards zero, whereas the reaction time to achieve net zero emissions is 30 years at best. Hence we might already have lost control of whether tipping happens".[35] In such a context of climate emergency, ecological liberalism, which allows each individual to choose their mitigation actions autonomously, is too permissive. If one is primarily concerned with preserving the basic liberties and wellbeing of individuals, it is actually necessary to rein in the excessive liberties some individuals and corporations have become used to in a context of economic deregulation to protect the rights and liberties of other individuals vulnerable to environmental impacts.

The second limitation is that these approaches do not sufficiently highlight the problems specific to technological measures. CDR, especially when deployed at large scale, poses major risks.[36] First, since many CDR technologies are yet unproven, they might eventually prove technologically, economically,

34 Steffen et al., "Trajectories."

35 Lenton et al., "Climate Tipping Points," 595.

36 See e.g., K. Dooley and S. Kartha, "Land-Based Negative Emissions: Risks for Climate Mitigation and Impacts on Sustainable Development," *International Environmental Agreements: Politics, Law and Economics* 18, no. 1 (2018).

or biophysically unfeasible. Second, their ecological and social costs might be unacceptable due to adverse impacts on biodiversity, food security, water resources, or human rights. Third, they might prove less effective than expected, due to reversal risks (human and natural actions disturbing reservoirs of carbon) or irreversible climate impacts during the overshoot periods caused by the crossing of tipping points. Fourth, CDR can start working as a substitute for immediate and ambitious emissions reductions, thereby increasing the risk of a large carbon budget overshoot.

Heyward is aware of (at least) some of these risks when she stresses that "such measures must be carried out in conjunction with programmes aimed at furthering global justice,"[37] but there is no guarantee that global justice is compatible with CDR. Given the many ethical issues raised by large-scale CDR, this focus on technological innovation requires a stronger justification. Just like betting on automatic population stabilization, betting on new technologies that would allow humanity to keep growing indefinitely is dangerous: as recent history shows, new technological innovations are as likely to increase GHG emissions (e.g., more carbon-intensive datacenters and new fracking methods) as they are to help decrease them (e.g., wind turbines and solar panels).[38]

Energy efficiency also raises problems because of the rebound effect. Energy and material efficiency gains in the production, supply, or use of a commodity or service regularly lead to increases in their use or consumption. Technical innovations such as increased energy efficiency can lead users to consume more of either the same product or other products, thereby cancelling out the energy gain achieved through increased efficiency.[39] The rebound effect is not systematic, and it is true that increased energy efficiency can lead to GHG emissions reductions, but this effect puts into question the idea that technological change can serve as a substitute for population policies.

Again, elements in the IPAT equation should not be considered as alternatives between which individuals can freely choose, but as complementary factors that both policies and individual behaviors should simultaneously incorporate.

37 Heyward, "A Growing Problem?" 732, n17.

38 Cafaro, "Climate Ethics and Population Policy."

39 F.-D. Vivien, "Effet-Rebond," in *Dictionnaire De La Pensée Écologique*, ed. D. Bourg and A. Papaux (PUF, 2015).

4 Incentive Policies

Which population policies should be promoted to fairly address climate change and biodiversity loss? As Hickey, Rieder, and Earl (2016, 853) helpfully suggest, population policies can be located on a "coercion spectrum," from policies with a low risk of coercion to policies with a high risk of coercion. While choice-enhancing measures such as improving education, promoting literacy, and improving women's economic opportunities are desirable for both intrinsic (providing access to basic goods) and instrumental (reducing GHG emissions and limiting biodiversity loss) reasons, coercive policies such as forced abortions and mandatory sterilizations represent unacceptable violations of the human rights to autonomy and bodily integrity. Educational and empowerment policies are certainly part of the toolbox of measures to reduce population growth, especially as they contribute not only to climate justice, but also social and global justice by protecting crucial rights, such as the right to education of girls and women. At the same time, we cannot count solely on such measures, as they can be too slow to change fertility rates or too ineffective to motivate enough people to have fewer children. Moreover, this kind of measure has more impact in developing countries than in developed ones: fertility rates and family planning needs are higher in Asia and Africa than in the EU and the US.[40]

Population policies can rely on incentives to complement choice-enhancing measures. Incentives are located in the mid-range of the coercion spectrum, meaning that they can be justified, on the crucial condition that they are properly designed and implemented. Incentives can be defined as a cost or benefit aimed at deliberately changing the behavior of an agent or group of agents. Regarding population growth reduction, negative incentives discourage parents from having too many children, while positive incentives encourage them to have fewer.

Incentives are compatible with democratic societies insofar as they allow parents to make their own choices, while at the same time economically encouraging them to have small families. It is however important to distinguish between justifiable and unjustifiable incentives. Three relevant criteria are: the purpose pursued by the authority setting incentives up; the margin incentives leave for a voluntary response; and the side effects on the people who are subject to incentives.[41]

40 C. Hickey, T. N. Rieder, and J. Earl, "Population Engineering and the Fight against Climate Change," *Social Theory and Practice* 42, no. 4 (2016), 855.

41 R. Grant, "Rethinking the Ethics of Incentives," *Journal of Economic Methodology* 22, no. 3 (2015), 360.

Worldwide, more than 162 million women have an unmet need for contraception, both in developed and developing countries.[42] They would like to plan their pregnancies and have smaller families, but are unable to do so due to the high cost of contraception, lack of access to contraceptives, lack of information about their use, or misinformation about their effects on health. Increasing the number of local clinics that provide free, relevant information to couples about contraceptive use can make a big difference. Similarly, making the contraceptive pill, the morning-after pill, and other contraceptive methods free of charge, as well as making optional certain time-consuming and costly doctor's appointments (especially for contraceptive pill renewals), are positive incentives that would greatly help women and couples who wish to better plan their family life. Overall, eliminating unintended pregnancies through contraceptive use could reduce the world's population by about 3 billion by 2100, which would lead to substantial reductions in global GHG emissions and protect biodiversity; in a social and global justice perspective, it would also allow women and couples to more freely pursue their life plan.[43]

Other positive incentives are possible, such as reducing taxes for households with small families. In many countries, such as France, Switzerland, or the US, a pro-natalist culture determines procreative incentives, especially through taxation.[44] The idea here is to reverse this incentive structure by encouraging couples to have fewer children. Making sterilization free for both men and women is also among the options to be considered. Giving a financial reward for sterilization would also be a possibility, but this measure should be implemented with caution. On the one hand, it is important not to set the amount of the reward too high, in order to avoid disadvantaged people reluctantly using it as an opportunity to obtain money. On the other hand, in a world of 8 billion people, facing dangerous climate change and mass species extinction, it is important not to set the amount of the reward too low, to avoid losing the incentivizing effect of the policy. An important point here is to respect the criterion of the margin left by incentives for a voluntary response. In general,

42 A. Haakenstad et al., "Measuring Contraceptive Method Mix, Prevalence, and Demand Satisfied by Age and Marital Status in 204 Countries and Territories, 1970–2019: A Systematic Analysis for the Global Burden of Disease Study 2019," *The Lancet* 400 (2022). This study focuses on unmet contraceptive needs for women; it is however also possible to calculate unmet needs by focusing on men or couples; see Stan Becker, "Measuring Unmet Need: Wives, Husbands or Couples?," *International Family Planning Perspectives* 25, no. 4 (1999).

43 Conly, *One Child*, 117.

44 This does not mean that in pro-natalist cultures people make children because of tax benefits; it simply means that, in such cultures, there is a stronger economic and social pressure, especially on women, to make children.

to be ethically acceptable, incentive policies must be implemented with particular attention to vulnerable groups, such as the poor, disabled people, or ethnic minorities.

Negative incentives are more difficult to justify, as they incur costs rather than provide benefits. Given the scale of the population problem and the gravity and urgency of climate change and biodiversity loss, these options should however also be explored. One possible measure here would be a tax increase for families with too many children. Since larger families benefit more from tax-financed public services, including public education, it seems fair that smaller families pay less tax and larger families pay more. This is not to say that parents alone should provide for the welfare of their children: the system of social solidarity established through progressive taxation is one of the most important achievements of contemporary democracies in terms of social justice. Rather, the aim is to achieve a fair tax system in which those who benefit the most pay the most—while obviously continuing to take account of different levels of income and wealth in a progressive taxation system, to avoid further disadvantaging the least well-off. The idea here is not to "punish" parents for having more children; the idea is precisely to avoid any form of punishment by using incentives rather than coercive regulation.

A major difficulty here is determining who has "too many" children. In the context of developed countries, Overall (2013, 183) and Rieder (2016, 37) support the moral duty not to give birth to more than two children per couple, or more than one child per person. Their objective is that each person should limit themselves to the procreative replacement of their person by a new person. Conly (2016) believes that more than one child per couple would already be too many. I see three strong reasons for supporting a one-child family norm in developed countries: (1) fertility rates are often already below two births per women; (2) procreation choices have a comparatively higher environmental impact; and (3) procreation means are more widely available and more affordable. The situation is different in poor countries with high fertility rates: supporting a one-child family norm in such a context would be much less feasible (given the limited effectiveness of government policies and availability and affordability of contraceptive means) and desirable (given possible tensions with human rights). The fertility rate reductions made possible by population policies will take more time to reach the objective of a one-child family in this context. There is therefore no "one size fits all" approach at the global level.

It is however important to keep in mind that population growth reduction or even population stabilization over the 21st century will probably not be sufficient to put an end to the transgression of planetary boundaries. If we take seriously the studies that stress the need to reduce the world's population to around 3 billion,[45] or even to at least 6.9 billion by 2100,[46] then population degrowth appears to be the best option. To avoid a dangerous anthropogenic interference with the climate system and preserve and restore biodiversity, policies that decrease the global human population have become necessary. This implies that choice-enhancing measures and positive and negative incentives should encourage one-child families as much as possible. Pro-natalist norms and incentive structures should be replaced by small family norms and incentive structures, both in developed and developing countries. Taking seriously the moral and political implications of the planetary boundary framework implies adopting a small family ethics and policy.

5 Defending Incentive Policies

5.1 *An Undue Inducement?*

Incentive population policies raise many objections. I focus here on those challenges that are the most relevant within the framework of this paper. A first objection supports that encouraging people to have fewer children may reduce parental autonomy and go against their interests. Incentives are morally problematic when the benefits they provide are large enough to short-circuit individuals' freedom of choice—this is the so-called "undue influence" argument.[47]

A first way to avoid this problem is to set the financial rewards at an amount that is not too high, as proposed above for the positive incentives for sterilization. But this would not be enough to avoid the objection altogether: even with a small amount, positive and negative incentives can limit the autonomy of agents by pushing them to have smaller families than they would like.

A second reply is that there is in fact already an incentive structure in place for reproductive choices. The pro-natalist culture of many developed and developing countries, as well as the institutions that support this culture,

45 Lianos and Pseiridis, "Sustainable Welfare and Optimum Population Size"; Tucker, "A Planet of 3 Billion"; Dasgupta, *Time and the Generations: Population Ethics for a Diminishing Planet*; Tamburino and Bravo, "Reconciling a Positive Ecological Balance with Human Development: A Quantitative Assessment."

46 Rogelj et al., "Scenarios Towards Limiting Global Mean Temperature Increase Below 1.5°C," 327; van Vuuren et al., "Alternative Pathways to the 1.5°C Target Reduce the Need for Negative Emission Technologies," 392.

47 Hickey, Rieder, and Earl, "Population Engineering and the Fight against Climate Change," 865–66.

already provide incentives for individuals to have children.[48] In North and South America, Europe, Australia, and even more so in the Middle East and sub-Saharan Africa, individuals' reproductive preferences are not autonomously chosen: religion, education, family environment, and social, tax, and health systems strongly encourage individuals to have children. Pro-natalist incentives such as tax reductions for larger families already exist, contributing to shaping individuals' reproductive choices.

The question, therefore, is not whether or not to incentivize, but how best to do so. In a context where population growth is contributing to the crossing of planetary boundaries, incentives for smaller families seem much more justified than incentives for larger families. All the more so if we take into account the fact that pronatalist policies are undermining the sexual and reproductive rights of women around the world, typically leading to abortion restrictions or prohibitions, and reduced access to sexual and reproductive healthcare.[49]

A third reply is that the kind of incentive policies supported in the previous section would not necessarily go against the interests of the parents. Empirical research shows that having fewer children, or not having any, does not reduce individuals' quality of life.[50] The life plan of many individuals includes starting a family, but people's life plans evolve over time, not only according to autonomous choices, but also through the influence of moral and social norms and public policies. If pro-natalists norms and policies are replaced by small family norms and policies, the life plan of many individuals will evolve accordingly, without necessarily interfering with their interests—provided that population policies are properly designed and implemented so that they do not violate their individual rights. Which leads us to a second objection.

5.2 *A Violation of the Individual Right to Procreate?*

The idea of procreative liberty is widely held. Robertson (1996, 24) expresses this idea in the following way: "Procreative liberty should enjoy presumptive primacy when conflicts about its exercise arise because control over whether one reproduces or not is central to personal identity, to dignity, and to the meaning of one's life." The individual right to procreate is commonly seen as a means of protecting this individual freedom. This right is defined in the United

48 T. Rieder, "Procreation, Adoption and the Contours of Obligation," *Journal of Applied Philosophy* 32, no. 3 (2015).

49 M. Scigliano, "Welcome to Gilead: Pronatalism and the Threat to Reproductive Rights," (Population Matters, 2021).

50 Hickey, Rieder, and Earl, "Population Engineering and the Fight against Climate Change," 866.

Nations Proclamation of Tehran as "a basic right [of the parents] to determine freely and responsibly the number and the spacing of their children".[51]

Raz (1984, 195) defines an individual right as follows: "'x has a right' if and only if x can have rights, and other things being equal, an aspect of x's well-being (his interest) is a sufficient reason for holding some other person(s) to be under a duty." Although having a child is not as fundamental an interest as someone's interest in health, subsistence, or life, for many people having a child is a very strong interest. Most people wish to have one or more children, and their conception of a good life largely relies on this possibility. The right to procreate protects, among other things, the interest in having biological offspring and the interest in having a family.[52] The desire to pass on one's genes and the desire to participate in family life are two bases of the right to procreate, and any state intervention that seeks to limit this right, whether in the form of educational, incentive, or coercive policies, is often viewed with suspicion.

But what exactly does the right to procreate imply? For example, can I claim it to oppose a tax increase caused by the birth of my second child? The two fundamental interests protected by this right do not imply an absolute right to freely determine the size of one's family. On the one hand, it is sufficient to have one child to transmit one's genetic heritage. Having more children certainly increases the guarantee that this heritage will be transmitted over several generations, but as in the case of most rights, no absolute guarantee is given. On the other hand, it is sufficient to have one child to found a family. A flourishing family is not necessarily a large family, but a family where the relationship between parent(s) and child(ren) is based on values such as love, respect, and gratitude. As Conly (2016, 49–52) stresses, the joys of family intimacy, the unique bonds between people, the shared story are all possible with just one child.

It is true that having a (biological or adopted) child deeply changes the people who become parents and the meaning of their lives. But this does not imply that with only one child, or no child at all, these people would have no meaning in their lives, no identity, or would necessarily be unhappy. People do have a right to procreate that protects fundamental interests, but these interests can be met with one child and are therefore not jeopardized by the kind of population policy supported here.

51 UN, "Final Act of the International Conference on Human Rights: Teheran, 22 April to 13 May 1968," ed. United Nations (1968), 4.

52 Conly, *Only One Child*, 39.

Like all liberties, procreative liberty is limited by the effects it can have on others. Incentive population policies partially interfere with people's reproductive autonomy, but this interference can be justified in a world where excessive human numbers are increasing the risk of abrupt climate change and leading to the extinction of numerous other species. On the one hand, these policies are not forcing individuals: they rather push them to have fewer children, while leaving them the possibility of having more. If individuals however choose to have a large family, they must pay more taxes to internalize the social and environmental costs of their choices (keeping in mind that the taxation scheme needs to be progressive to take into account inequality difference between income classes). On the other hand, even if individual choices are influenced, this small restriction on parental autonomy may also be justified because it is intended to protect others from their reproductive choices. Sticking to a conception of freedom according to which the individual has reproductive rights but no reproductive responsibilities could amount to tolerating multiple injustices, particularly climate injustices and interspecies injustices.[53]

5.3 *More Children for People with Ecological Awareness?*

A third objection focuses on the values transmitted by parents to their children. Population policies push people with a strong ecological awareness to have fewer children. This means there is a risk that the offspring of people with such awareness will be fewer in number than the offspring of people with no or low ecological awareness. According to this objection, values such as respect for the environment, love of nature, and mindfulness are transmitted from parents to children. For this reason, people with high ecological awareness should have more children than those with no or low ecological awareness to counterbalance the effects of people who are not concerned about the environment.[54]

A first response to this argument is that such behavior would contradict the values that ecologically aware parents uphold and seek to pass on to their children.[55] If they want their behaviors to reflect their ethical commitments, ecologically aware people should not have multiple children, as this action contradicts their objective to reduce their overall ecological footprint.

A second response is that the argument is based on the problematic assumption that "children adopt many of the beliefs and practices of their parents".[56]

53 On the notion of "interspecies injustice", see Wienhues (2020).

54 S. Wisor, "Is There a Moral Obligation to Limit Family Size?" *Philosophy & Public Policy Quarterly* 29, no. 3/4 (2009): 29.

55 Overall, *Why Have Children? The Ethical Debate*, 190–91.

56 Wisor, "Is There a Moral Obligation to Limit Family Size?" 29.

There is actually no guarantee for this. On the contrary: growing up in a large family will likely lead children to doubt their parents' ecological awareness. If parents try to justify their contradictory behavior, their children will probably see it as a manifestation of hypocrisy, as "do as I say, but not as I do."

The third reply draws on virtue ethics. A person who is aware of the existence of planetary boundaries and of the considerable pressure of human activities on these boundaries would not take Earth system science data lightly and would no doubt feel some responsibility. While they may well choose to have one biological child, maybe two, it is unlikely that they would choose to have more. They would probably see it as environmentally vicious behavior, showing a lack of consideration for the planet and for others. The virtue of temperance, for example, encourages people to live in relative moderation, to reduce their consumption, and to look after their carbon footprint. Given the excess of consumption and the additional GHG emissions inevitably caused by a new child, a temperate person would likely choose to have a small family and would be more likely to accept population policies.[57]

6 Conclusion

While population growth is not the sole cause of climate change and biodiversity loss, it contributes significantly to the increasing environmental impact of human societies. However, influential philosophical approaches continue to neglect or at least downplay the importance of the "P" component of the IPAT equation. The core claim of this chapter is that that all three components of the IPAT equation need to be addressed urgently and simultaneously. There is no justification to further avoid addressing the topic of population reduction, no matter how unpopular and controversial this topic can be. A central part of the argument for addressing population is that there is already a pronatalist incentive structure in place; that is, institutions already encourage individuals to make certain reproductive choices. There are compelling reasons in our

57 Rieder, *Toward a Small Family Ethic*, 58. Note that the first two responses to the objection also rely to some extent on virtue ethics. They draw on the virtue of integrity, which requires that we integrate our deep ethical commitments into our everyday actions. Another issue here is the one of measuring ecological awareness. One could argue that one's ecological footprint could serve as a proxy for one's degree of ecological awareness, but people with high ecological awareness may have very little control over their carbon footprint, just like people with low ecological awareness may have a small carbon footprint for other reasons. Virtue ethics allows to avoid this difficulty by focusing on the kind of motivation that drives people to have children or not to have them.

context of climate disruption and biodiversity loss to encourage individuals to have fewer children. Furthermore, the right to procreate is not absolute; like all rights, it is limited by the impact that individual choices have on others. Additionally, if properly designed and applied, population policies could not only promote climate justice, but also social and global justice, by protecting the rights and freedoms of women.

Increased population and affluence are two of the main drivers of climate change and biodiversity loss.[58] Taking the planetary boundary framework seriously implies not only reducing population, but also reducing economic growth.[59] Technological innovation and managerial efficiency will not be enough to avoid dangerous anthropogenic interference with the climate system; by themselves, they certainly cannot end massive global biodiversity loss.[60] Avoiding dangerous climate change and a mass species extinction involves economic changes accompanied by ethically acceptable population policies. Sufficiency in affluence and technology are not sufficient; we also need sufficiency in population. Reducing excessive wealth and replacing unsustainable technologies with cleaner ones are certainly important measures to address climate change and biodiversity loss, but these measures will remain insufficient as long as they are not complemented by measures to reach a more sustainable global population size.

For this reason, it would be relevant for future research to articulate the IPAT equation with limitarianism, the distributive justice approach according to which "it is not morally permissible to have more resources than are needed to fully flourish in life".[61] Whereas sufficientarianism sets a lower threshold on the quantity of goods people should have to live a decent or flourishing life, limitarianism sets an upper threshold on the goods individuals should be allowed to possess. This approach would not only further justify the need to put an absolute limit on extreme wealth, linking this form of wealth with global poverty and inequalities, historical injustices, and threats to democracy.[62] It would also allow governments implementing population policies and facing a

58 IPCC, "Summary for Policymakers"; Cafaro, Hansson, and Götmark, "Overpopulation Is a Major Cause of Biodiversity Loss."

59 E. Crist et al., "Protecting Half the Planet and Transforming Human Systems Are Complementary Goals," *Frontiers in Conservation Science* 2 (2021); H. Daly and J. Farley, *Ecological Economics, Second Edition: Principles and Applications* (Island Press, 2011).

60 W. Ripple et al., "World Scientists' Warning of a Climate Emergency 2022," *BioScience* 72, no. 12 (2022).

61 I. Robeyns, "Having Too Much," in *NOMOS LVI: Wealth. Yearbook of the American Society for Political and Legal Philosophy*, edited by J. Knight and M. Schwartzberg (New York University Press, 2017), 1.

62 I. Robeyns, *Limitarianism: The Case Against Extreme Wealth* (Astra House, 2024).

declining tax base to generate a massive alternative source of revenue and restore their fiscal agency, as limitarian institutions impose a confiscatory tax on the fortunes of the super-rich and put an end to tax avoidance and tax evasion. A fewer number of taxpayers does not necessarily mean fewer social services: a more progressive taxation scheme can compensate for population reductions. Systematically integrating the IPAT equation into a limitarian approach, in particular by articulating wealth limitarianism with ecological limitarianism, is a relevant topic for future research.

Acknowledgements

This chapter is an adapted, revised, and updated version of Bourban (2019). I am very grateful to Philip Cafaro and Lisa Broussois for their highly valuable feedback on previous drafts of the paper. I would also like to thank the editors of the volume as well as the audience of the Human Tech Colloquium at the University of Twente in October 2022 for their helpful comments and questions.

References

Armstrong McKay, D., et al. (2022). "Exceeding 1.5°C global warming could trigger multiple climate tipping points." *Science* 377(6611):eabn7950. doi: doi:10.1126/science.abn7950.

Arnsperger, C. and Bourg, D. (2017). Écologie intégrale. Pour une société permacirculaire. Paris PUF.

Becker, S. (1999). "Measuring Unmet Need: Wives, Husbands or Couples?" *International Family Planning Perspectives* 25(4):172–180. doi: 10.2307/2991881.

Bourban, M. (2019). "Croissance démographique et changement climatique : repenser nos politiques dans le cadre des limites planétaires." *La Pensée écologique* 3(1):19–37. doi: https://doi.org/10.3917/lpe.003.0019.

Bourban, M. (2022). "Ethics, Energy Transition, and Ecological Citizenship." In *Comprehensive Renewable Energy (Second Edition)*, edited by T. Letcher, 204–220. Elsevier.

Bourban, M. (2023). "Mitigation Duties." In *Handbook of the Philosophy of Climate Change*, edited by G. Pellegrino and M. Di Paola, 721–758. Springer International Publishing.

Cafaro, P. (2011). "Beyond business as usual: alternative wedges to avoid catastrophic climate change and create sustainable societies." In *The Ethics of Global Climate Change*, edited by D. Arnold, 192–215. Cambridge University Press.

Cafaro, P. (2012). "Climate ethics and population policy." *WIREs Climate Change* 3(1): 45–61. doi: 10.1002/wcc.153.

Cafaro, P. (2022). "Climate ethics and population policy: A review of recent philosophical work." *WIREs Climate Change* 13(2):e748. doi: https://doi.org/10.1002/wcc.748.

Cafaro, P., Hansson P., and Götmark, F. (2022). "Overpopulation is a major cause of biodiversity loss and smaller human populations are necessary to preserve what is left." *Biological Conservation* 272:109646. doi: https://doi.org/10.1016/j.biocon.2022.109646.

Caney, S. (2020). "Human Rights, Population, and Climate Change." In *Human Rights and 21st Century Challenges: Poverty, Conflict, and the Environment*, edited by D. Akande, et al., 348–369. Oxford University Press.

Cohen, J. (2010). "Population and Climate Change." *Proceedings of the American Philosophical Society* 154(2):158–182.

Conly, S. (2016). *One Child: Do We Have a Right to More?* Oxford University Press.

Crist, E., et al. (2021). "Protecting Half the Planet and Transforming Human Systems Are Complementary Goals." *Frontiers in Conservation Science* 2. doi: 10.3389/fcosc.2021.761292.

Daly, H. and Farley, J. (2011). *Ecological Economics, Second Edition: Principles and Applications.* Island Press.

Dasgupta, P. (2019). *Time and the Generations: Population Ethics for a Diminishing Planet.* Columbia University Press.

Dhakal, S., et al. (2022). "Emissions Trends and Drivers." In *IPCC, 2022: Climate Change 2022: Mitigation of Climate Change. Contribution of Working Group III to the Sixth Assessment Report of the Intergovernmental Panel on Climate Change*, edited by P. R. Shukla, et al. Cambridge University Press.

Dodson, J., et al. (2020). "Population growth and climate change: Addressing the overlooked threat multiplier." *Science of The Total Environment* 748:141346. doi: https://doi.org/10.1016/j.scitotenv.2020.141346.

Dooley, K. and Kartha, S. (2018). "Land-based negative emissions: risks for climate mitigation and impacts on sustainable development." *International Environmental Agreements: Politics, Law and Economics* 18(1):79–98. doi: 10.1007/s10784-017-9382-9.

Engelman, R. and Johnson, D. (2019). "Removing barriers to family planning, empowering sustainable environmental conservation: A background paper and call for action." In *Margaret Pyke Trust.*

Fragnière, A. (2016). "Climate change and individual duties." *WIREs Climate Change* 7(6):798–814. doi: 10.1002/wcc.422.

Gardiner, S. (2011). *A Perfect Moral Storm: The Ethical Tragedy of Climate Change.* Oxford University Press.

Grant, R. (2015). "Rethinking the ethics of incentives." *Journal of Economic Methodology* 22(3):354–372. doi: 10.1080/1350178X.2015.1071506.

Haakenstad, A., et al. (2022). "Measuring contraceptive method mix, prevalence, and demand satisfied by age and marital status in 204 countries and territories, 1970–2019: A systematic analysis for the Global Burden of Disease Study 2019." *The Lancet* 400:295–327. doi: 10.1016/S0140-6736(22)00936-9.

Heyward, C. (2012). "A Growing Problem? Dealing with Population Increases in Climate Justice." *Ethical Perspectives* 19(4): 703–732.

Hickey, C., Rieder, T., and Earl, J. (2016). "Population Engineering and the Fight against Climate Change." *Social Theory and Practice* 42(4): 845–870. doi: 10.5840/soctheorpract201642430.

IPBES. (2019). Summary for policymakers of the global assessment report on biodiversity and ecosystem services (summary for policy makers). IPBES Plenary at its seventh session.

IPCC. (2014). "Summary for Policymakers." In *Climate Change 2014: Mitigation of Climate Change. Contribution of Working Group III to the Fifth Assessment Report of the Intergovernmental Panel on Climate Change*, edited by O. Edenhofer, et al., 1–30. Cambridge University Press.

IPCC. (2022). "Summary for Policymakers." In *Climate Change 2022: Mitigation of Climate Change. Contribution of Working Group III to the Sixth Assessment Report of the Intergovernmental Panel on Climate Change*, edited by P. R. Shukla, et al., 7–53. Cambridge University Press.

Jiang, L. and Hardee, K. (2011). "How do Recent Population Trends Matter to Climate Change?" *Population Research and Policy Review* 30(2): 287–312. doi: 10.1007/s11113-010-9189-7.

Laurent, É. (2020). *The New Environmental Economics: Sustainability and Justice*. Polity.

Lenton, T., et al. (2019). "Climate tipping points—too risky to bet against." *Nature* 575:592–595. doi: 10.1038/d41586-019-03595-0.

Lianos, T. and Pseiridis, A. (2016). "Sustainable welfare and optimum population size." *Environment, Development and Sustainability* 18(6): 1679–1699. doi: 10.1007/s10668-015-9711-5.

Murtaugh, P. and Schlax, M. (2009). "Reproduction and the carbon legacies of individuals." *Global Environmental Change* 19(1):14–20. doi: 10.1016/j.gloenvcha.2008.10.007.

O'Neill, B., et al. (2010). "Global demographic trends and future carbon emissions." *Proceedings of the National Academy of Sciences* 107(41): 17521–17526. doi: 10.1073/pnas.1004581107.

O'Sullivan, J. (2018). "Synergy between Population Policy, Climate Adaptation and Mitigation." In *Pathways to a Sustainable Economy: Bridging the Gap between Paris Climate Change Commitments and Net Zero Emissions*, edited by M. Hossain, R. Hales and T. Sarker, 103–125. Springer International Publishing.

OurWorldInData. (2022). "Per capita consumption-based CO_2 emissions, 2020." accessed 21.11.2022. https://ourworldindata.org/grapher/consumption-co_2-per-capita.

Overall, C. (2013). *Why Have Children? The Ethical Debate.* MIT Press.

Oxfam. (2015). *Extreme Carbon Inequality.* Oxfam International.

Pacala, S. and Socolow, R. (2004). "Stabilization Wedges: Solving the Climate Problem for the Next 50 Years with Current Technologies." *Science* 305(5686): 968–972. doi: doi:10.1126/science.1100103.

Parfit, D. (1986). *Reasons and Persons.* Oxford University Press.

Persson, L., et al. (2022). "Outside the Safe Operating Space of the Planetary Boundary for Novel Entities." *Environmental Science & Technology.* doi: 10.1021/acs.est.1c04158.

Raz, J. (1984). "On the Nature of Rights." *Mind* 93(370): 194–214.

Richardson, J., et al. (2023). "Earth beyond six of nine Planetary Boundaries." *Science Advances* 9(37):eadh2458. doi: 10.1126/sciadv.adh2458.

Rieder, T. (2015). "Procreation, Adoption and the Contours of Obligation." *Journal of Applied Philosophy* 32(3): 293–309. doi: 10.1111/japp.12099.

Rieder, T. (2016). *Toward a Small Family Ethic: How Overpopulation and Climate Change Are Affecting the Morality of Procreation.* Springer.

Ripple, W. et al. (2022). "World Scientists' Warning of a Climate Emergency 2022." *Bio-Science* 72(12): 1149–1155. doi: 10.1093/biosci/biac083.

Ritchie, H. and Roser, M. (2019). "CO_2 and Greenhouse Gas Emissions." https://ourworldindata.org/co_2-and-other-greenhouse-gas-emissions.

Robertson, J. (1996). *Children of Choice: Freedom and the New Reproductive Technologies.* Princeton University Press.

Robeyns, I. (2017). "Having Too Much." In *NOMOS LVI: Wealth. Yearbook of the American Society for Political and Legal Philosophy,* edited by J. Knight and M. Schwartzberg. New York University Press.

Robeyns, I. (2024). *Limitarianism: The Case Against Extreme Wealth.* Astra House.

Robinson, W. and Ross, J. (2007). *The Global Family Planning Revolution.* The World Bank.

Rogelj, J., et al. (2018). "Scenarios towards limiting global mean temperature increase below 1.5°C." *Nature Climate Change* 8(4): 325–332. doi: 10.1038/s41558-018-0091-3.

Scigliano, M. (2021). *Welcome to Gilead: Pronatalism and the Threat to Reproductive Rights.* Population Matters.

Steffen, W., et al. (2015). "Planetary boundaries: Guiding human development on a changing planet." *Science* 347(6223): 1259855. doi: 10.1126/science.1259855.

Steffen, W., et al. (2018). "Trajectories of the Earth System in the Anthropocene." *Proceedings of the National Academy of Sciences* 115(33): 8252–8259. doi: 10.1073/pnas.1810141115.

Tamburino, L. and Bravo, G. (2021). "Reconciling a positive ecological balance with human development: A quantitative assessment." *Ecological Indicators* 129: 107973. doi: https://doi.org/10.1016/j.ecolind.2021.107973.

Tucker, C. (2019). "A planet of 3 billion." *Atlas Observatory.*

UN. (1968). Final Act of the International Conference on Human Rights: Teheran, 22 April to 13 May 1968, edited by United Nations.

UN. (2022). *World Population Prospects 2022: Summary of Results*. UN DESA.

Van Vuuren, D., et al. (2018). "Alternative pathways to the 1.5°C target reduce the need for negative emission technologies." *Nature Climate Change* 8(5): 391–397. doi: 10.1038/s41558-018-0119-8.

Vivien, F.-D. (2015). "Effet-rebond." In *Dictionnaire de la pensée écologique*, edited by D. Bourg and A. Papaux, 382–384. PUF.

Wang-Erlandsson, L. et al. (2022). "A planetary boundary for green water." *Nature Reviews Earth & Environment* 3(6): 380–392. doi: 10.1038/s43017-022-00287-8.

Wienhues, Anna. (2020). *Ecological Justice and the Extinction Crisis: Giving Living Beings their Due*. 1 ed: Bristol University Press.

Wisor, S. (2009). "Is There a Moral Obligation to Limit Family Size?" *Philosophy & Public Policy Quarterly* 29(3/4): 89–117.

Wynes, S. and Nicholas, K. (2017). "The climate mitigation gap: education and government recommendations miss the most effective individual actions." *Environmental Research Letters* 12(7): 1–9.

CHAPTER 11

Sufficient Population? Checklist for a More Proper Discourse on the Degrowth of Human Numbers in Affluent Societies

Pasi Heikkurinen and Joonas Uotinen

1 Introduction

It is a broadly accepted fact that the number of humans on Earth contributes to the ecological footprint of human societies.[1] The more humans there are—and will be—on this planet, the greater the environmental impact is.[2] We also know that the population discourse tends to cause discomfort due to its historical connotations to events reducing selected "ethnic" human populaces, like the Holocaust, the Holodomor in the Soviet Union, or Cambodian genocide. It goes without saying that any anti-humanistic thoughts and actions are a major threat even in contemporary societies. Ignoring the question of population, however, will arguably not also be the proper reaction to deal with this major factor of human impact on the environment. This assumption leads us to ask: how we should talk about the unsustainability of the continuing population growth in the affluent societies, mainly located in the northern hemisphere? We extend the question to include also stabilized highly populated contexts in affluent countries, which are beyond the biocapacity of a given region. On the global scale, the stabilization of population around 10 billion inhabitants is also considered highly questionable from a sustainability point of view.

The degrowth movement effectively addresses the problem of economic growth but is silent on the problem of population growth.[3] Owing to its stark realization about the impossibility of infinite growth on a finite planet, would

1 For example, Wackernagel, M., and Rees, W. *Our Ecological Footprint* (Vol. 9). (New Society Publishers, 1998); Costanza, R. (2000) "The dynamics of the ecological footprint concept," *Ecological Economics* 32(3); Fanning, A., et al. (2022) "The social shortfall and ecological overshoot of nations," *Nature Sustainability* 5(1).

2 Crist, E., et al. (2022) "Scientists' warning on population," *Science of The Total Environment* 845.

3 For example, Latouche, S. ([2007] 2009) *Farewell to Growth*. Polity Press; Kallis, G., et al. (2018) "Research on degrowth," *Annual Review of Environment and Resources* 43; Kerschner, C., et al. (2018) "Degrowth and technology," *Journal of Cleaner Production* 197.

 | DOI:10.1163/9789004746404_012

it not also be meaningful to talk about degrowth in relation to population? Could we imagine that sufficiency—as a focal component in the degrowth movement—is needed also in the number of humans, be that on the scale of households, communities, villages, cities, countries, and the globe? And if so, what do we need to consider when engaging in the depopulation discourse to stay ethically inclusive and avoid reproducing unjust and repressive language, as well as to steer away from violent practices? While we identify the sensitivity of the population discourse, we find it problematic for sustainability scholarship at large, in case it is left outside the debate and policy. After all, the most significant environmental problems are caused by the sheer size of the global human population and its affluence, and as technology has not delivered its promise of decoupling environmental harm from human numbers and affluence, our increasing numbers and material wealth have led to the ecological carrying capacity of the Earth being exceeded.[4]

In this chapter, we will use Finland as a case to discuss the relationship between population growth and ecological sustainability. The main motivation for the inquiry's focus on the affluent society context comes from our own place of dwelling, which is Finland, and then secondly, because of Finland's high per person consumption rate.[5] The reduction of the number of high consuming people is of course ecologically more effective in comparison to reducing the amount of low consuming people. In this chapter, first, we claim that it is important to spark a discussion on the way we could address the degrowth of human populace both globally and locally in a manner that is not only effective from an ecological point of view but also socially inclusive.

Secondly, for this difficult undertaking of addressing the unsustainability of increasing human numbers, the chapter proposes a checklist of five discursive starting points. By complementing the common focus on the amount of people and their consumption levels, we shall consider the quality of human life not limited to—or defined by—its material conditions. We suggest that contentment aligns with a sufficient number of human numbers. Sufficient population, including the amount of reproduction and the number of people

4 For example, Barnosky, A. D., et al. (2012) "Approaching a state shift in Earth's biosphere," *Nature* 486 (7401); Ceballos, G., et al. (2015) "Accelerated modern human-induced species losses," *Science Advances* 1 (5); Cook, J. et al. (2016) "Consensus on consensus," *Environmental Research Letters* 11 (4).

5 Using Finland as the case signifies our analysis on the population discourse to locate on the nation state, by which we do not imply that this level of analysis is ultimately the most appropriate one. In the task of arriving at socially and ecologically just discourse, i.e., a more proper population discourse, we need to employ also other levels of analysis, not least the scale of households—the more affluent and less affluent families.

(kept) alive, is largely defined by cultural factors at issue. In its simplest, both "too few" and "too many" humans in a given context reduces contentment, as well as it can be speculated to be an outcome of a lost contentment. That is, there must be something fundamentally wrong in a society that either over-reproduces or under-reproduces its members. By underpopulation we refer to a hypothetical situation where humans are extremely few in numbers, jeopardizing the continuity of that species. This is considered an issue not only from the human point of view, but also from the perspective of biodiversity, since losing a species always reduces the diversity of life.

2 Foundational Premises

Informed by sustainability sciences, the most effective way to improve global ecological sustainability would be to reduce the amount of affluence and slow down the growth of human populace. While the problem of wealth—in relation to sustainability—is increasingly being addressed (albeit still not adequately), the issue of population largely remains a taboo in sustainability studies and policy. Calls for "economic degrowth" we may already hear even in mainstream newspapers and public debate, but only a few must have heard of "population degrowth." One obvious reason is the growth hegemony, which links the affluence of an economy to its amount of people. Population growth is often perceived desired from an economic viewpoint—more people in the marketplace signifies more potential for consumption and taxation, and more people in the labor force enhance the dependency ratio in the economy.[6] Population degrowth again creates economic challenges for societies based on growth.[7] If growth is preferred, neither economic or population degrowth is desired.

The discursive difficulty of addressing the amount of population is even accentuated in the affluent societies of the Global North due to their connection to the colonial past.[8] After all the highest rates of population growth are in

6 A better dependency ratio means that the nation has a better ratio of those in the labor force (aged 15 to 64, for example) to those outside the labor force (aged 0 to 14 and 65 +, for example).

7 Jones, C. (2022) "The end of economic growth?" *American Economic Review* 112(11).

8 This surely is the case in some countries more than in others, if one for example compares the histories of the UK and Finland, the differences are clear. And again, this is not to say that Finland has no colonial past (see Nyyssönen, 2013), but rather to pinpoint the difference in degree and kind.

the so-called developing countries,[9] which signifies that population discourse in affluent societies touches on the rights of not only "the developed" but also the "developing." Even to use the word of "population growth," not to mention "population degrowth," can be misinterpreted as a misuse of power of rich old white men (from the north) to interfere with the lives of poor young black women (from the south). In other words, population discourse on a global scale might be understood to automatically signify claims about the need to lower population in developing societies, which of course does not have to be the case. The unequal power structures, however, are something we must take seriously when examining the question of population. Let us proceed building the checklist for a more proper population discourse that is not blind to *unequal power structures*.

Another related major pitfall in talking about population growth in any critical fashion comes from a myopic and anthropocentric understanding of human rights. In 1979, The Convention on the Elimination of All Forms of Discrimination against Women (CEDAW) defined reproduction as basic right of all couples and individuals: everyone is "to decide freely and responsibly the number, spacing and timing of their children and to have the information and means to do so, and the right to attain the highest standard of sexual and reproductive health."[10] What if we were to extend these rights to cover also future human generations? Then anything weakening their ability to exercise similar rights should not be off the table? Also, for the moment, the ways of exercising the right have deteriorated and will continue to deteriorate the ability of non-human species to satisfy their needs to the extent that we are on the brink of a new mass extinction wave.[11] Do not the non-humans have rights,[12] and what would they be like? How should we approach the right today if it jeopardizes similar rights of future humanity and the ability to live of non-human species today and in the future? Is the contemporary way of exercising the right "responsible" as stated in the CEDAW formulation?

---

9 For a critical treatment of the term 'development', please see Escobar (1992), inter alia.

10 UN (1979/1981) Convention on the Elimination of All Forms of Discrimination against Women. Adopted and opened for signature, ratification and accession by General Assembly resolution 34/180 of 18 December 1979. Entry into force 3 September 1981, in accordance with article 27(1).

11 For example, Barnosky, A. et al. (2011) "Has the Earth's sixth mass extinction already arrived?" *Nature* 471 (7336).

12 For example, Thiel, P. and Hallgren, H. (2018) "Rights of Nature as a Prerequisite for Sustainability," in *Strongly Sustainable Societies*, edited by K. Bonnedahl and P. Heikkurinen. Routledge.

Additionally, the amount of human population can be, and already largely is, affected without (directly) interfering on the right of self-determination of individuals. As individuals', as well as couples', decisions are always done within the framework of opportunities, responsibilities and conventions provided by societies, policies are already influencing the human numbers, even if they are seldom explicated in public. Indeed, political discourses rarely connect to the nexus of population and the environment. The question of 'how to approach' this nexus remains. Let us proceed to building the checklist for a more proper population discourse that is not *anti-human rights*.

These two rather basic but foundational premises of a more proper population discourse, namely (1) acknowledging the unequal power structures in society (and seeking to do something about them) and (2) respecting the human rights also in an intergenerational framing, will be our guiding normative assumptions in this chapter. They will also form the groundwork for addressing the question of how we should talk about population sizes. That is, we posit that a more appropriate discourse on population degrowth, as well as *the search for* such a discourse, shall begin from becoming aware of the past, present, and future of the unequal power structures, including the colonial past of many affluent countries, and secondly, it shall be respectful towards the UN human rights declaration. Since these two are very dangerous, yet rather common pitfalls; colonialism and anti-humanism, why even bother to ask about population growth? What could be more important than equity and human life today? Let us look at Finland as a case study to understand the problem.

3 Case Finland, Embedded in the Biosphere

Both geographical Finland and Finnish culture are part of the biosphere, the sum of all ecosystems. Biosphere refers specifically to the parts of the Earth where life is possible. Life can be broadly defined as a situation in which something is born and alive and not yet dead.[13] However, the definition of life is difficult with respect to its temporal and local boundaries, as well as due to

13 What we mean here by 'dead' or 'death' is a very mundane and ordinary conceptualization of it. It is not to refer to a complete disappearance of a matter-energetic entity, like a human being annihilation from the cycle of life, but rather to a transformation of the entity's state of being. For instance, the body's particles return to the ground or to the atmosphere, and become part of the circulation of coal, which gives birth to new forms of life. In perception, one may seem to disappear, as this matter-energetic transformation is so significant—and hereby changing the manner in which we (those alive) perceive 'the dead.'

the diversity and interdependencies of the phenomenon called "life." It surely is difficult to also define "death", but what is important here is to acknowledge the common preference of "those alive" to not "die." And although we cannot give an exact definition of life (or death), it is true in the experience of many—we feel alive.

The lives of human individuals, as well as groups of people and cultures have always been in constant danger. Survival struggles are constantly fought around the world. Communities are born, they live, and they die. Over the last hundred years, however, the struggle for life has reached global proportions. Studies by leading biologists and paleontologists indicate that we live amid the sixth wave of mass extinction.[14] What makes this species loss exceptional is at least two things. The first is the scientific fact that extinctions are largely the product of a single species—*homo sapiens*. Another peculiar point is that, although human communities have had extensive information available about the ongoing collapse, its causes and consequences, the necessary measures to slow down or stop species' deaths have not been taken.

Life on Earth is based on the interdependence of biotic and abiotic processes (such as species and ecosystems).[15] Consequently, without a diverse spectrum of species, the conditions for the continuity of human life will also deteriorate. This means that the progress of planetary mass extinction will arguably advance also human extinction. In addition to human altruistic interests, we may thus be concerned about the life conditions of other species as the collapsing biosphere also threatens ourselves.[16]

What factors then would allow the continuity of life on Earth? This issue is addressed in the field of sustainability studies, which examines the links between the environment, society, and the economy. Scholars in the field aim to increase understanding of the unsustainability phenomenon and provide means for sustainable change. This field of research also offers a view of the factors that constitute human impact on the environment.[17] According to the so-called IPAT equation, the three most important factors influencing the environment (I = Impact) are population (P = Population), wealth (A = Affluence)

14 Wake, D. and Vredenburg, V. (2008) "Colloquium Paper: Are We in the Midst of the Sixth Mass Extinction?" *Proceedings of the National Academy of Sciences of the United States of America* 105 Suppl 1; Barnosky et al., "Has the Earth's Sixth Mass Extinction Arrived Already?"; Ceballos et al. "Accelerated Modern Human-Induced Species Losses".

15 Capra, F., *The Web of Life* (Anchor, 1996); Heikkurinen, P., et al. (2016) "Organising in the Anthropocene," *Journal of Cleaner Production* 113.

16 Barnosky A. et al. (2012), "Approaching a state shift in the Earth's biosphere," *Nature* 486(7401); Ehrlich, P., and Ehrlich, A. (2013) "Can a Collapse of Global Civilization Be Avoided?" *Proceedings Biological Sciences* 280 (1754).

17 Bonnedahl K, and Heikkurinen, P. *Strongly Sustainable Societies*. (Routledge, 2019).

and technology (T = Technology).[18] In the equation, the human impact on the environment (i.e., the so-called ecological footprint) thus consists of the total number of people, their material standard of living (or consumption) and the environmental impact of the technology they build and use. The first two factors, population and wealth, are considered as root problems of unsustainability. The third factor, on the other hand, is often seen as a means to mitigate the effects of the first two, through making production and consumption more ecologically efficient with technological improvements.

There are also critical views on the mitigating role of technology.[19] Namely, it is noteworthy that despite the efficiencies achieved through technology, the global total consumption and production have not declined, but have rather continued to grow.[20] The efficiency achievements that have been met are of an order of magnitude that do not correspond to the disadvantages of population and wealth. Locally reported benefits, such as the Finnish government's reduced carbon emissions, are largely due to a territorial emissions accounting, in which case products and services produced and purchased elsewhere are excluded from the calculation.[21] There is a lot of recent research literature on the problems of technology, particularly concerning its failure to decouple the economic and population growth from environmental damage.[22] In the context of Finnish social policy, Helne, Hirvilammi and Laatu have shown how technology is not decoupling economic growth from its environmental impacts.[23]

Affluence and the degradation of the biosphere, in turn, have been linked in scholarly debates, via income level[24] and consumption.[25] That is, wealth—

18 Holdren, J., and Ehrlich, P. (1974) "Human population and the global environment," *American Scientist* 62 (3); Chertow, M. (2000) "The IPAT Equation and Its Variants," *Journal of Industrial Ecology* 4 (4); Holdren, J. (2018) "Brief History of IPAT," *The Journal of Population and Sustainability* 2 (2).

19 For example, Heikkurinen, P. (2018) "Degrowth by means of technology? A treatise for an ethos of releasement," *Journal of Cleaner Production* 197; Ruuska, T., Heikkurinen, P., Levasseur, T., and Gare, A. (2024). Redefining Violence for the Anthropocene: From Ecocide to Ecological Civilization. *Cosmos and History: The Journal of Natural and Social Philosophy*, 20(2), 56–76.

20 IEA (2018) "Global energy & CO_2 status report 2017". International Energy Agency.

21 Barrett, J., et al. (2013) "Consumption-based GHG emission accounting: a UK case study," *Climate Policy* 13 (4).

22 For example, Wiedmann et al. 2015; Heikkurinen 2018; Giampietro 2019; Hickel & Kallis 2019; Parrique et al. 2019.

23 Helne, T., Hirvilammi, T., Laatu, M. (2012) *Sosiaalipolitiikka rajallisella maapallolla*. Kelan tutkimusosasto; see also Bonnedahl and Heikkurinen 2019.

24 Ulvila, M., and Pasanen, J. (2009) *Sustainable Futures: Replacing Growth Imperative and Hierarchies with Sustainable Ways*. Ministry of Foreign Affairs of Finland; Ulvila & Wilén 2017.

25 Brown, P. M., Cameron, L. D. (2000) "What can be done to reduce overconsumption?" *Ecological Economics* 32 (1); Lorek & Fuchs 2013; Bonnedahl & Caramujo 2018.

measured by income level and consumption—is conceptualized as the primary culprit for humans' environmental impact. Also, the quantitative growth of production—including the ideology of productivism—has been criticized in sustainability research.[26] The problem of growing wealth has also emerged in the general debate on consumption reduction efforts, e.g., through the Finnish media,[27] environmental organizations,[28] and the international degrowth movement. Of these three IPAT factors, population is clearly the least addressed.

All the research gaps aside, the utmost importance of discoursing population is its matter-energetic consequences on life, to both human and non-human. As the social and economic sustainability depend on ecological sustainability (referring particularly to the state of the biosphere that allows for the continuity of a diverse life), environmental impact of humans is too vital to be left out of the debate. Without ecological sustainability, no society, economy, organization, or individual can be sustainable, that is, a unit or system that sustains a diverse life. All human constructions, such as equity and rights, are in a subordinate position to the biosphere, including life itself. This assumption of *embedding all human activity within the biospher*e raises a question of how the national and local levels in Finland should respond to global challenges, such as the global demographic explosion that followed the Industrial Revolution (Figure 11.1). The global demographic development now incorporates an increase of more than 130 million people annually.[29] This is the troubling "hockey stick" curve.

As the Figure 11.1 demonstrates, in the beginning of the 19th century, the world's human population was one billion, now it is 8 billion. This is a significant addition to the ecological footprint of the human species. A radical decline in living standards can remedy this undesirable development to an extent, as could the right kind of technological organization (in theory), but it is clear that every new human life means an increase in the collective environmental

26 Bonnedahl, K. J., Eriksson, J. (2007) "Sustainable economic organisation: simply a matter of reconceptualisation or a need for a new ethics?" *International Journal of Innovation and Sustainable Development* 2 (1); Heikkurinen & Bonnedahl 2013; Heikkurinen et al., 2021.

27 E.g. Yle, "Kulutuksen vähentäminen auttaa ympäristöä" (In English: reducing consumption helps the environment), 27.9.2007, https://yle.fi/aihe/artikkeli/2007/09/27/kulutuksen-vahentaminen-auttaa-ymparistoa. Kansan uutiset, "Ylikulutus uhkaa terveyttä – suomalaiset syövät maapallon 3,6-kertaisesti" (In English: overconsumption threatens health – the Finns eat 3.6 earths), 1.8.2018, https://www.kansanuutiset.fi/artikkeli/3939119-ylikulutus-uhkaa-terveytta-suomalaiset-syovat-maapallon-36-kertaisesti.

28 E.g. WWF, "Ylikulutuksen vaikutukset" (consequences of overconsumption), 2019, https://wwf.fi/uhat/ylikulutus/.

29 The World Bank, "Health, Nutrition and Population: Population Dashboard," accessed in 2019, http://datatopics.worldbank.org/health/population.

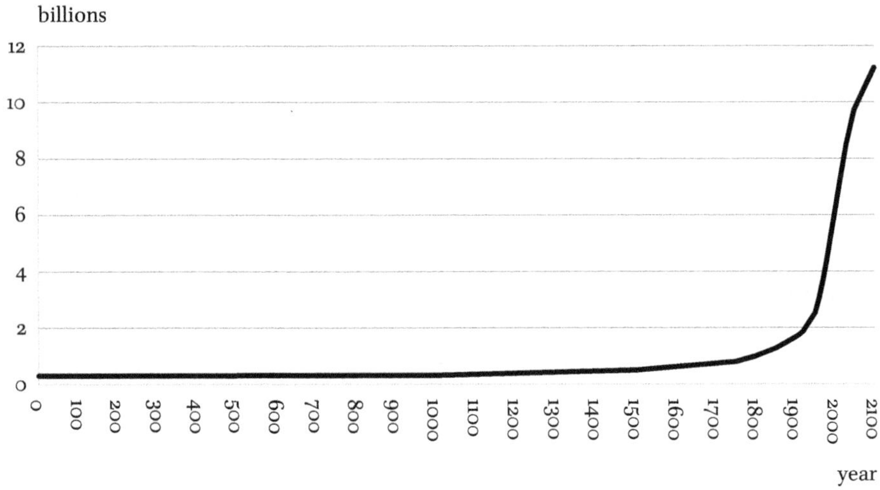

FIGURE 11.1 The number of humans on Earth from 0 to 2100 AD

impact of the human species. In this context, it is worth noting that the ecological footprints of different existing and nascent human individuals (as well as cultures) can differ significantly from the average individual. According to the IPAT formula, it largely depends on the level of affluence of the unit under consideration. In other words, the more materially wealthy the unit in question (like an individual or a society) is, the more it consumes natural resources and produces harmful emissions, such as greenhouse gases, and therefore the larger its ecological footprint is. To put it simply: if an example community A lives 10 times more prosperously than community B, then community B can have 10 times more people and yet the ecological footprints of A and B remain the same (*ceteris paribus*). Thus, the greatest environmental damage is caused by communities and societies with a large number of materially prosperous long-lived people.

The rich may find it difficult to accept this fact because of the possibility of guilt and shame, as well as because consumption is reckoned to be a human right.[30] One attempt to circumvent this correlation between economic capital and environmental damage is to think that 'although I am prosperous, I, firstly, do not consume all my wealth' and, secondly, 'I consume it sustainably.' However, this thinking is based on two erroneous assumptions. The first is the

30 For example, Miller, J. C., and Stovall, T. (2019) "The "right to consume"? Re-thinking the dynamics of exclusion/inclusion in consumer society," *Consumption Markets & Culture* 22 (5–6).

illusion that wealth could exist without it being used (either on its own or by someone else) or that the accumulated wealth would not have had a negative impact on the environment at the time of its acquisition. In the current, global market-oriented and growth-based economic system all wealth deposited in banks is in constant use, for example. It finances both the bank's own investments and the additional loans made by its customers. In other words, in the Finnish monetary system—which is of course largely a subordinate to those decisions made at the Economic and Monetary Union of the European Union (which again is a part of the global market), it is extremely difficult to own anything without it being used to grow the economy. If you deposit your salary or own a residential apartment with mortgage, then this wealth or mortgage of yours is indirectly consumed through financial instruments of various sorts.

The other misconception is about the myth of sustainable consumption. Qualitative improvements in products and services can only have a limited effect on the quantity of the consumed products and services. This is due to the so-called rebound effect, also known as the Jevons paradox,[31] which explains how the "saved" environmental benefit due to a higher quality purchase decision, becomes consumed elsewhere. In other words, we can eat vegetarian food, drive an electric car, move around in public places, and live in an energy-efficient house, but, to the extent that these can be done more cheaply, our total consumption as quantity of goods or services consumed may increase reducing or reverting gains in our total ecological footprint. The biosphere cares little about *where* the resources it provides are used. From the perspective of the continuation of life, the priority is on *how much* and *how fast* the resources are used. And this, of course, is influenced by how many humans—or consumers in the context of an economy—are using them. The utopia of intangible consumption, where societies could maintain a high standard of living by shifting from consumption of products to services, is very much linked to this misconception. Unfortunately (we suppose), however, no human activity has been successfully disconnected from its material-energetic base. In other words, while one kind of consumption may press the ecology less than others, evidence of so-called "intangible services" in its absolute sense is lacking. Humans as matter-energetic entities are embedded in the biosphere.

The Finnish policy debate is concerned about the number of people, but for a reason that is mainly economic, i.e., aimed at maintaining and increasing economic growth rather than seeking ecological sustainability. Policy makers in Finland want population growth! It is thought that by having more healthy

31 Jevons, W. S. (1906) *The Coal Question.* Macmillan; Alcott, B. (2005) "Jevons' paradox," *Ecological Economics* 54 (1).

FIGURE 11.2 The number of live births in Finland since 1900

entrants into the labor force, by immigration or fertility rates, the dependency ratio is better, and the welfare state is easier to fund. Overall, a shrinking market is undesired as it may lead to a weaker competitive position and less affluence to share. A degrowing population in Finland may deteriorate the prospects for economic growth.

And if the size of the economy does not grow, then a growth-based society may find it difficult to raise or even maintain current living standards. The decline in the birth rate in Finland is a central cause to the concerns about the change in the demographic structure and the related insolvency of the pension system. If the birth rate continues to decline (Figure 11.2), the economic outlook is worrying in Finland. What will happen to the prospering Finland if it cannot keep up with international economic competition?

If the birth rate does continue to fall, as Figure 11.2 suggests, and the standard of living in Finland is challenged, the future looks promising from the viewpoint of ecological sustainability. A growth-based economy constantly needs more human and non-human resources, and if these are not available, it may be that the system will stall.[32] The consequences are uncertain and short-term disadvantages would probably occur for humans, but every slowdown

32 Ruuska, T. (2017) *Reproduction of Capitalism in the 21st Century: Higher Education and Ecological Crisis*. Ph.D. dissertation, Aalto University.

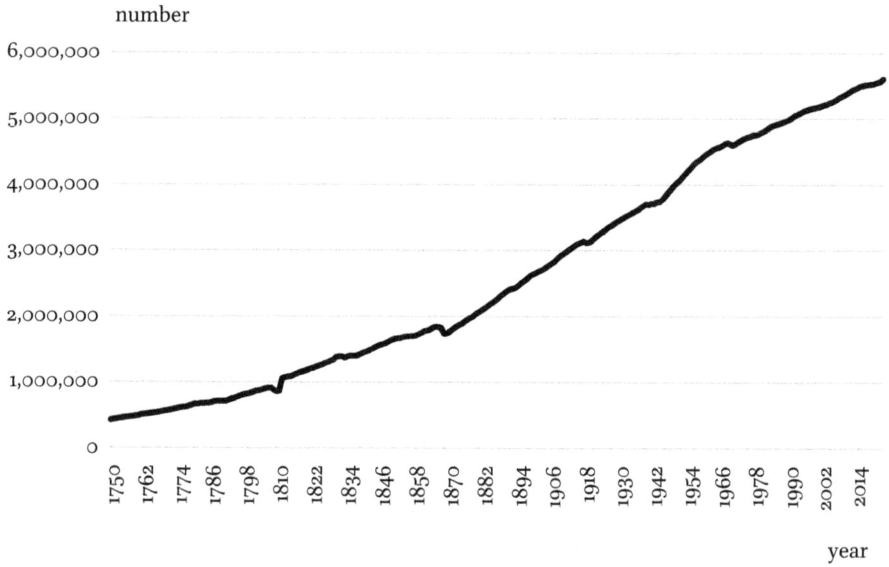

FIGURE 11.3 Population growth in Finland since 1750

and industry outage will give the biosphere time to recover.[33] However, the impact of this on global ecological sustainability would be small, as the share of the Finnish society in the world economy is tiny, although per capita consumption in Finland is at the forefront of the world. Furthermore, the population of Finland continues to grow and the decline caused by the birth rate will hardly be visible in the number of Finns for a long time to come (Figure 11.3). A main reason for population growth in Finland is immigration—paradoxically partly caused by global environmental change.

Owing to immigration, the economic concern on lowering the birth rate of Finns (Figure 11.2) is less pertinent. If Finland will partake in global flows—in which money and workers move relatively freely—there will be enough consumers and producers to keep the economy growing. It seems that these humans, however, will not be native Finns. From a very narrow, xenophobic national identity or ethnic perspective, where one may want to preserve the so-called Finnish genetic and cultural heritage, this may be seen as a problem, but the economy will hardly be affected by the "mixing". As long as the economy keeps growing, the capitalist system is probably not bothered. Of course, the Finnish economy would benefit from an increase in its own population

33 Ruuska, T., Heikkurinen, P., Levasseur, T., and Gare, A. (2024). Redefining Violence for the Anthropocene: From Ecocide to Ecological Civilization. *Cosmos and History: The Journal of Natural and Social Philosophy*, 20(2), 56–76.

as well, but if that lost is compensated by immigrants, no economic problem arises. From the ecological sustainability point of view, however, in this era of excessive consumption and reproduction, decline in birth rates enhances the conditions for the continuation of a diverse life.

4 A Checklist for Population Discourse

What kind of lessons can be drawn from the case and its analysis so far? First, it is noteworthy that the biosphere is disinterested in the ethnic origin and other demographic factors of its destroyers. For the planet, it makes little difference if the produced battery is consumed in conjunction with a kid's toy or her dad's electric car.

In other words, it is irrelevant from the viewpoint of ecological sustainability, who consumes resources and where. From the viewpoint of social sustainability, this issue is surely important. The more evenly resources and living standards are distributed over time and space, the better are the chances for social justice to be realized and social order preserved. However, it must be remembered that every social system is subordinate to the biosphere. In practice this means that even if we distribute all affluence completely evenly among all humans (and repeat it if resources were to accumulate unevenly again) the problem of over-consumption will not go away, unfortunately. Despite this, and in other words, the over-consuming part of humanity and their prosperous societies—such as Finland—should at the same time reduce their own total consumption, while distributing wealth more evenly. Though it is of utmost significance from the viewpoint of ecological sustainability that starting to decline the overall standard of living to something that is "sufficient", from the viewpoint of social sustainability it is important that the already existing wealth is more evenly distributed. In any case, for the global consumption to reach a level of sufficiency, the standard of living in affluent societies must decrease sharply. Gross Domestic Product (GDP) has been found to correlate closely with the ecological footprint[34] and can still measure the desired direction of national economies, though inversely. With a Gross Domestic Product (at current market prices) of 234.5€ billion in 2018, representing 42,504€ per capita,[35] Finland certainly belongs to these affluent societies that must change their course.

Even the dystopian decoupling option, in which the economy is declining in terms of GDP, but the ecological footprint is not shrinking, cannot be ruled

34 Daly, H. (1996) *Beyond Growth*. Beacon Press; Victor 2008; Jackson 2009; Latouche 2009.

35 Data from Statistics Finland, 2019, https://www.stat.fi/.

TABLE 11.1 Five-point checklist for a more proper discourse on sufficient population

Point	Description	Elaboration through negation
1st	Population discourse should be inclusive by engaging actors from the private, public and third sectors	Population discourse should not be only for the individual to deliberate
2nd	Population discourse should be transparent by openly disclosing its means and ends	Population discourse should not be built on hidden agendas
3rd	Population discourse should be advancing human rights by including the capabilities of future generations to exercise the same rights in the analysis	Population discourse should not reject the capabilities of the present humans
4th	Population discourse should be critical towards technological solutions by emphasizing non-anthropocentrism	Population discourse should not be limited to the analysis of humans
5th	Population discourse should be emphasizing contentment within limits	Population discourse should not be merely about the number of human offspring

out as one of the scenarios for the future. Also, if GDP degrowth would take place, it is unclear how population growth would be affected. So far it appears that higher affluence has been linked to lower population growth. Does it then follow that lower affluence will lead to higher population growth again? It may be that even with degrowth in terms of affluence, the need for population degrowth cannot be circumvented. All the arguments presented so far call for the necessity of including population growth into the discussions on sustainability. They beg for a closer look on the possibilities for social change in relation to population growth. We suggest a checklist of five key points for discussing the population question.

4.1 *Population Discourse Should Be Inclusive by Engaging Actors from the Private, Public and Third Sectors* (*Not Only for the Individual to Deliberate*)

In addition to statistical projections, there is a need to look at the enablers of social change in different sectors of society to reach sustainability.[36] Public

---

36 Bonnedahl, K. J. and Heikkurinen, P. (eds.) (2019) *Strongly Sustainable Societies: Organising Human Activities on a Hot and Full Earth*. Routledge; Heikkurinen et al. 2019.

actors, like legislators, operate largely based on a democratic logic, while private actors, such as companies and households, tend to operate through the logic of the market. In the third sector, civil society sets standards based on socio-cultural development. In practice, the drivers of change do not neatly fall into any of these three sectors alone, but common problems are defined and solved through politics and laws as well as norms and market forces. It would therefore be absurd to argue that something closely linked to sustainability, such as the population issue, should be addressed in only one area of society. Just as the problem of wealth cannot be left to consumers and businesses alone, the question of population cannot be ruled out from the public debate. Yet, the problems of population growth *from the viewpoint of sustainability* are often disregarded and left to the responsibility of the individuals.

4.2 *Population Discourse Should Be Transparent by Openly Disclosing Its Means and Ends (Not Building on Hidden Agendas)*

For example, journalist David Roberts lists the reasons why he has chosen not to talk about population growth in the context of environmental problems, even though he acknowledges the link between population growth and the ecological footprint.[37] He argues that by focusing on different instruments (such as girls' education), population growth can be slowed down and even put on a degrowth track without talking about it. This is some kind of a hidden agenda for policy that may, on the one hand, reduce moral and political friction, and on the other hand, prevent the transparency of moral and political action. The secreted approach is probably politically amenable, but at the same time the idea that we can solve the demographic problem without talking about it seems suspicious. There are also unsolved ethical issues in social engineering.[38]

The risks that Roberts and many others seek to avoid with their hidden agenda on tackling population growth are noteworthy and serious concerns. In his writing, he links the population issue to 'racism, xenophobia, and racial breeding'. According to him, talk about population growth means the desire to reduce the number of certain population groups instead of others. And this is the kind of population discourse we most certainly want to avoid, as already outlined in the beginning of this chapter. We nevertheless want to challenge

37 Roberts, D. (2018) "I'm an environmental journalist, but I never write about overpopulation. Here's why." *Vox*, 29 November 2018. https://www.vox.com/energy-and-environment/2017/9/26/16356524/the-population-question.

38 Mouton, F., et al. (2015) "Necessity for ethics in social engineering research," *Computers & Security*, 55.

any straightforward causal connections (even correlation) between using the term population growth in the context of environmental problems and racism, xenophobia, and racial breeding. Arguably there is no such shared purpose among those who engage in the talk about population growth. At least for us there is not. Population discourse should not be understood to signify claims about the need to lower population with an unethical ideology or by any unethical means. In fact, leaving the painful things out of the public and academic discussion sounds more problematic to us, than engaging in these discourses. In Finland, only a decade ago it was difficult to talk about economic growth in any critical fashion because it was seen to lead to communism. Now some are unfairly associating the population discourse to fascism and Nazism.

Of course, Roberts' opinions are probably driven by an altruistic concern for justice among humans, as well as by some form of awakening to the human species needing a biosphere to continue its existence (and not destroy more others). However, many who speak about population may have a (more or less) selfish fear of being stigmatized. Especially in certain social contexts, norms sharply forbid talking about it. By saying the P-word out loud, one can indeed be labeled a racist, racial hygienist, xenophobic, neo-Malthusian, Nazi, or fascist. The fear of stigmatization is therefore real and will certainly explain in part why the public debate on population growth is non-existent. How should we then talk about population growth? To reiterate the rather obvious guideline: by avoiding any statements that can be interpreted in a way that is likely to lead to further stigmatization. This of course is difficult to do as in the present age the discussion culture is rather polarized.

4.3 *Population Discourse Should Be Advancing Human Rights by Including the Capabilities of Future Generations in the Analysis (Not Rejecting the Capabilities of the Present Humans)*

Another major hindrance for talking about population growth in any critical fashion comes from a largely shared understanding of human rights combined with a misconception that population policies need to necessarily curb them.[39] As Kopnina and Washington discuss, however, there are many non-coercive policies that support reduction of population growth that actually ameliorate the women's possibilities to realize their own wishes for reproduction. These are to guarantee secondary school for all including girls, ensuring gender equality throughout the societies and making contraceptives accessible to all among others. Furthermore, there may be other ways to affect population

39 Kopnina, H. and Washington, H. (2016) "Discussing Why Population Growth Is Still Ignored or Denied," *Chinese Journal of Population Resources and Environment* 14 (2).

growth without touching the right of self-determination. As individuals' decisions are always done within the framework of opportunities, responsibilities and conventions provided by societies, policies already influence the amount of population even if they are not explicitly about it.

This is evidenced by lower fertility in more affluent countries.[40] For example, the abolition of child labor and the implementation of compulsory education combined with individualism, that removes the moral duty of taking care of one's old parents, seems to make children largely an economic burden instead of an asset. While these examples, particularly the former, may not be those decrees one should tamper with, they attest to the individuals' choices being largely done in the context of the societal structures that are formed in part through policies.

A follow-up question may need to be asked: now that we know the undesired environmental impacts of population growth, should we not also investigate human rights in the light of sustainability studies? If we extend the rights to cover also future human generations, then anything weakening their actualizable possibilities—capabilities[41]—to exercise similar rights to have children, for example, should not be off the table. Economic insecurities have been argued to be related to low fertility even in the affluent Finland.[42] The looming environmental collapse in part caused by the population numbers will cause economic insecurities and, therefore, will interfere with the capability of future generations to have children. Nevertheless, for answering the question of how we should talk about population, we deduce that a way that advances rather than challenges human rights will be successful.

4.4 *Population Discourse Should Be Critical towards Technological Solutions by Emphasizing Non-anthropocentrism (Not Limiting the Analysis to Humans)*

The widespread techno-optimism in current environmental policy also supports the silence around population. It is safe to immerse oneself in the utopian idea that humans—and especially affluent societies such as Finland—will always be able to solve the problems it faces by means of technology. Although there is no empirical evidence for the improvement of the state of the biosphere except very locally, techno-optimistic thinking is still very popular in affluent societies. An acclaimed exception to this is the success in stopping

40 For example, Oishi and Diener 2014.

41 Sen, A. (1999) *Development as Freedom*. Knopf.

42 Jalovaara, M., et al. (2019) "Education, Gender, and Cohort Fertility in the Nordic Countries," *European Journal of Population*, 35(3).

the destruction of the Ozone layer accomplished through international agreements, such as the Montreal Protocol in the late 1980s,[43] but these policy solutions are not actually "technological innovation."

The techno-optimist thinking is probably popular in part because by believing in technical solutions, individuals and communities do not need to radically change their current thoughts or actions. On the other hand, in the field of economic policy, there is little critical talk on population growth, except in the opposite sense where this is not enough growth. And those who are critical towards economic growth, the degrowth movement and/or Sufficiency Movement, do not want to talk about population degrowth, because the capitalist economic system is seen to increase consumption so much that population growth is not considered a relevant factor. This of course is rather absurd. As IPAT advises us, both population and affluence, as well as technology, matter for human environmental impact.

Sustainability debates often take "the ecological" and "the social" as interchangeable or substitutable domains of life. In contrast to "strong sustainability", this is called "weak sustainability".[44] At worst, some hubristic thinking even subordinates the biosphere to human life. That is, instead of seeing human wellbeing as dependent on ecological sustainability, anthropocentric thought considers the humankind capable of transcending the matter-energetic base on life. By placing the human species above other earthbound species, this kind of thinking represents inter-species racism or human supremacist ideas.[45] The population discourse, that considers both the non-humans and future human generations of equal value with the present humans, does not justify those actions found to be harmful to other species and future humans.

4.5 *Population Discourse Should Be Emphasizing Contentment within Limits* (*Not Merely the Number of Human Offspring*)

Contentment—understood as satisfaction with life, frequent pleasant emotions, infrequent unpleasant emotions, and experiencing life as meaningful—may align to a large extent with what humans desire, making it a promising avenue forward. Inspired by the sustainable wellbeing literature,[46] we could approach the question of population by asking how humans could live good

43 Newman, P. et al. (2009) "What would have happened to the ozone layer if chlorofluorocarbons (CFCs) had not been regulated?" *Atmospheric Chemistry and Physics* 9(6).

44 For example, Bonnedahl and Heikkurinen 2019.

45 Jensen, D. (2016) *The myth of human supremacy*. Seven Stories Press; Crist, E. (2017) "The affliction of human supremacy," *The Ecological Citizen* 1(1).

46 Raworth, K. (2012) "A Safe and Just Space for Humanity: Can We Live within the Doughnut," *Oxfam Discussion Paper*; Helne et al. 2012; Hirvilammi and Helne 2014; Gough 2017.

lives in their local contexts and cultures, while not exceeding the ecological limits? Or conversely, how could humans not transgress ecological boundaries and still have contentment? Offspring, or their absence, and their cultural meanings play a significant part in earthlings' lives making them a natural point of departure for this question.

While a comprehensive literature review is beyond the scope of this article, some results can be presented on the connection between reproduction and contentment. Studies are somewhat contradictory and dependent on methodologies and terminologies used, but at least according to time use investigations, parenting tends to relate to higher life satisfaction, positive emotions, and meaningfulness, though may also have burdens such as increased time-stress.[47] Becoming a parent is found to associate with increased wellbeing; but the gains vanish with time;[48] and consequent children at least beyond 3rd and 4th child may already depress wellbeing in comparison to the pre-parenting levels.[49] The latter result is particularly interesting to Finland as it was from a relatively robust twin-study in another Nordic country, Denmark.

Associations between parenting and contentment appear to differ between individual and cross-regional levels, however. In the cross-national level, higher fertility rates (average number of children born per woman in the case that all women lived through the ages when they can have children) relate to lower life satisfaction.[50] However, societies with higher fertility rates appear to experience their lives more meaningful. The relation between meaningfulness and the number of children is positive (and statistically significant), and remains (weakly) statistically significant after controlling for religion and GDP per capita.[51]

These findings suggest that at least in some set of societies, wellbeing gains from offspring may be highest for the first children and when becoming a parent; but may vanish in the longer term and with a higher number of children. However, it is noticeable that independent, adult children may have a wellbeing

47 Nelson-Coffey, S. K. (2018) "Married ... with Children: The Science of Well-Being in Marriage and Family Life," in *Handbook of Well-Being*, edited by Diener, Oishi, and Tay. DEF Publishers.

48 Nelson-Coffey, "Married ... with Children"; Kohler, H., Behrman, J. R., Skytthe, A. (2005) "Partner + children = happiness? The effects of partnerships and fertility on well-being," *Population and development review* 31 (3).

49 Kohler, H., Behrman, J. R., and Skytthe, A. (2005) "Partner + children = happiness? The effects of partnerships and fertility on well-being," *Population and development review* 31(3): 407–45.

50 Oishi, S. and Diener, E. (2014) "Residents of Poor Nations Have a Greater Sense of Meaning in Life Than Residents of Wealthy Nations," *Psychological Science* 25 (2).

51 Ibid.

increasing impact.[52] Although these results suggest that having few children instead of many might be the most wellbeing enhancing; it is unclear to what extent these parenting-wellbeing results represent only affluent societies.

These results indeed appear to match well the low fertility in the so-called developed countries. If several children do not contribute to contentment here, but do so in the developing countries, the relation between number of children and contentment could depend on societal structures and norms. In such a case, it appears that contentment would also then closely align with human choices as people in developed countries do make fewer children. Not only parents' contentment is relevant, however, but also that of the children, and the generations to come. Overall, either these results provide hope in that few children are better than many; or in that also contentment and human individual choice related to children are sensitive to societal structures and norms. In the latter case, population degrowth might not need to relate to lower contentment in affluent societies.

5 Conclusion

From a materialist non-anthropocentric point of view, the population question is largely a matter of how space and resources are divided between the earthlings. If a species has too little space and resources to live, then some kind of a good life is impossible or at least extremely difficult. Thus, in the name of inter-species justice, it appears clear that the material standard of human living must fall considerably to reach a state of sufficiency, both in Finland and globally. The more humans we ask this Earth to host, the more drastic the reduction in affluence must be. This is another inconvenient natural scientific truth.

On the one hand, wealth (measured in GDP and income levels, as well as consumption) is the single most important factor from the viewpoint of ecological sustainability, as it can be affected quickly without inhumane solutions. On the other hand, according to the IPAT formula: if the birth rate and the population are reduced, then there is no need to reduce the standard of living as much. The most effective and long-term way is to consider both factors (P + A).

52 Shields, M. and Wooden, M. (2003) "Marriage, children and subjective well-being," an unpublished article presented at the 8th Australian Institute of Family Studies Conference. https://melbourneinstitute.unimelb.edu.au/assets/documents/hilda-bibliography/conference-papers-lectures/2003/Shields_etal_Marriage,_Children_and_Subjective_Well-being.pdf.

But before any further practical plans (e.g. policy) to the population question can be proposed in a constructive tone, the discourse needs to develop a more shared understanding of the causes of the ecological footprint. For instance, is the IPAT agreed on? If not, what explains the human environmental impact if not population, affluence, and technology?

It seems that the neglect of discussing unsustainability is being phased out in Finland, but the next challenge is to remove the neglect on its causes. Nowadays, we seem to have fewer climate change deniers, for example, but still very many who deny the causes of climate change. In this chapter, we have proposed a checklist of five key points to get humans in the private, public and third sectors to discuss the problems of population growth. Moreover, we suggest that if the human environmental impact factors put forward by the international scientific community are not accepted as a basis for discussion, both national and global political dialogue and measures to ensure the continuity of life will more likely fail. If there is an inadequately shared common ground on the problem, the proposed solutions will sound absurd.[53] In addition to natural science (such as disciplines of ecology and biology), findings of social science and humanities (such as anthropology, sociology, and economics) should be taken into account when formulating problem descriptions in order to make contemporary diagnosis as inclusive as possible. Furthermore, the discussion on population in affluent societies, as well as elsewhere, cannot be limited to scientists but must host a variety of societal actors and sections.

Contentment may promise a politically feasible approach to discuss the population question especially if it is combined with noncoercive policies and equality considerations—the incentives created by policies should not make the poor worse off in terms of reproduction either. So far, the evidence is somewhat mixed, but sufficiency in population or population degrowth may not need to entail lower contentment at least in all its dimensions. For sustainability, Finnish population policy is now at the center of social, economic and environmental policy. The grand challenge for the next decade will be to find ways to discuss difficult topics, like the need for population degrowth, not population growth. This includes weighting the local and global development benefits and costs from a variety of perspectives. This chapter has added one perspective to this debate from Finland and will hopefully contribute to a more proper discourse of the nexus of human numbers and environmental harm.

53 For example, Wynes, S. and Nicholas, K. A. (2017) "The climate mitigation gap: education and government recommendations miss the most effective individual actions," *Environmental Research Letters* 12 (7).

Acknowledgements

We want to express our thanks to the editors of this volume, Toni Ruuska and Tina Nyfors, for your comments and editing work. We also want to thank our colleagues at the Universities of Helsinki and Turku; as well as at The Finnish Institute for Health and Welfare, particularly Maria Vaalavuo for her insightful remarks. The first version of this chapter was presented at the 4th Peaceful Coexistence Colloquium in 2021. We thank the organizers and participants for the great discussions on the theme of sufficiency. This work was supported by the Academy of Finland: SustAgeable Consortium Project 'Economic and Social Sustainability Across Time and Space in an Ageing Society' (345384); and Research Project 'Skills of Self-Provisioning in Rural Communities' (343277).

References

Alcott, B. (2005) "Jevons' paradox," *Ecological Economics* 54(1): 9–21. https://doi.org/10.1016/j.ecolecon.2005.03.020.

Barnosky, A. D., et al. (2012) "Approaching a state shift in Earth's biosphere," *Nature* 486(7401): 52–58. https://doi.org/10.1038/nature11018.

Barnosky, A. D., et al. (2011) "Has the Earth's sixth mass extinction already arrived?" *Nature* 471(7336): 51–57. https://doi.org/10.1038/nature09678.

Barrett, J., et al. (2013) "Consumption-based GHG emission accounting: a UK case study," *Climate Policy* 13(4): 451–470. https://doi.org/10.1080/14693062.2013.788858.

Bonnedahl, K. J. and Eriksson, J. (2007) "Sustainable economic organisation: Simply a matter of reconceptualisation or a need for a new ethics?" *International Journal of Innovation and Sustainable Development* 2(1): 97. https://doi.org/10.1504/ijisd.2007.016060.

Bonnedahl, K. J. and Caramujo, M. J. (2018) "Beyond an absolving role for sustainable development: Assessing consumption as a basis for sustainable societies," *Sustainable Development* 27(1): 61–68. https://doi.org/10.1002/sd.1862.

Bonnedahl, K. J. and Heikkurinen, P. (eds.) (2019) *Strongly Sustainable Societies: Organising Human Activities on a Hot and Full Earth*. Routledge. https://doi.org/10.26530/oapen_1000439.

Brown, P. M., and Cameron, L. D. (2000) "What can be done to reduce overconsumption?" *Ecological Economics* 32(1): 27–41. https://doi.org/10.1016/s0921-8009(99)00093-2.

Capra, F. (1996) *The Web of Life: A New Scientific Understanding of Living Systems*. Anchor.

Ceballos, G., et al. (2015) "Accelerated Modern Human-Induced Species Losses: Entering the Sixth Mass Extinction," *Science Advances* 1(5): e1400253–e1400253. https://doi.org/10.1126/sciadv.1400253.

Chertow, M. R. (2000) "The IPAT Equation and Its Variants," *Journal of Industrial Ecology* 4(4): 13–29. https://doi.org/10.1162/10881980052541927.

Cook, J., et al. (2016) "Consensus on consensus: A synthesis of consensus estimates on human-caused global warming," *Environmental Research Letters* 11(4): 048002.

Costanza, R. (2000) "The dynamics of the ecological footprint concept," *Ecological economics* 32(3): 341–345. https://doi.org/10.1016/S0921-8009(99)00150-0.

Crist, E. (2017) "The affliction of human supremacy," *The Ecological Citizen* 1(1): 61–64. https://www.ecologicalcitizen.net/pdfs/v01n1-11.pdf

Crist, E., et al. (2022) "Scientists' warning on population," *Science of The Total Environment* 845: 157166. https://doi.org/10.1016/j.scitotenv.2022.157166

Daly, H. E. (1996) *Beyond Growth: The Economics of Sustainable Development.* Beacon Press.

Ehrlich, P. R and Ehrlich, A. H. (2013) "Can a Collapse of Global Civilization Be Avoided?" *Proceedings Biological Sciences* 280(1754): 20122845–20122845. https://doi.org/10.1098/rspb.2012.2845.

Escobar, A. (1992) "Imagining a post-development era? Critical thought, development and social movements," *Social Text* 31/32: 20–56.

Fanning, A. L., O'Neill, D. W., Hickel, J., Roux, N. (2022) "The social shortfall and ecological overshoot of nations," *Nature Sustainability* 5(1): 26–36. https://doi.org/10.1038/s41893-021-00799-z.

Giampietro, M. (2019) "On the Circular Bioeconomy and Decoupling: Implications for Sustainable Growth," *Ecological Economics* 162: 143–56. https://doi.org/10.1016/j.ecolecon.2019.05.001.

Gough, I. (2017) *Heat, Greed and Human Need: Climate Change, Capitalism and Sustainable Wellbeing*. Edward Elgar Publishing.

Heikkurinen, P. (2018) "Degrowth by means of technology? A treatise for an ethos of releasement," *Journal of Cleaner Production* 197: 1654–65. https://doi.org/10.1016/j.jclepro.2016.07.070.

Heikkurinen, P., Ruuska, T., Kuokkanen, A., Russell, S. (2021) "Leaving productivism behind: Towards a holistic and processual philosophy of ecological management," *Philosophy of Management* 20(1): 21–36. https://doi.org/10.1007/s40926-019-00109-w

Heikkurinen, P., Young, C. W., Morgan, E. (2019) "Business for sustainable change: Extending eco-efficiency and eco-sufficiency strategies to consumers," *Journal of Cleaner Production* 218: 656–664. https://doi.org/10.1016/j.jclepro.2019.02.053

Heikkurinen, P. and Bonnedahl, K. J. (2013) "Corporate responsibility for sustainable development: a review and conceptual comparison of market – and stakeholder -oriented strategies," *Journal of Cleaner Production* 43: 191–98. https://doi.org/10.1016/j.jclepro.2012.12.021.

Heikkurinen, P., et al. (2016) "Organising in the Anthropocene: an ontological outline for ecocentric theorizing," *Journal of Cleaner Production* 113: 705–14. https://doi.org/10.1016/j.jclepro.2015.12.016.

Helne, T., Hirvilammi, T., Laatu, M. (2012) *Sosiaalipolitiikka rajallisella maapallolla*. Kelan tutkimusosasto.

Hickel, J. and Kallis, G. (2019) "Is Green Growth Possible?" *New Political Economy* 25(4): 469–86. https://doi.org/10.1080/13563467.2019.1598964.

Hirvilammi, T. and Helne, T. (2014) "Changing paradigms: A sketch for sustainable well-being and ecosocial policy," *Sustainability* 6(4): 2160–2175. https://doi.org/10.3390/su6042160

Holdren, J. P. and Ehrlich, P. R. (1974) "Human population and the global environment: Population growth, rising per capita material consumption, and disruptive technologies have made civilization a global ecological force," *American Scientist* 62(3): 282–92.

Holdren, J. (2018) "Brief History of IPAT," *The Journal of Population and Sustainability* 2(2): 66–74. https://doi.org/10.3197/jps.2018.2.2.66.

IEA (2018) "Global energy & CO_2 status report 2017". International Energy Agency. https://www.iea.org/reports/global-energy-co$_2$-status-report-2017.

Jackson, T. (2009) *Prosperity without Growth: Economics for a Finite Planet*. Earthscan.

Jalovaara, M., et al. (2019) "Education, Gender, and Cohort Fertility in the Nordic Countries," *European Journal of Population*, 35(3): 563–586. https://doi.org/10.1007/s10680-018-9492-2

Jensen, D. (2016) *The Myth of Human Supremacy*. Seven Stories Press.

Jevons, W. S. (1906) *The Coal Question: An inquiry concerning the progress of the nation, and the probable exhaustion of our coal-mines*. Macmillan.

Jones, C. I. (2022) "The end of economic growth? Unintended consequences of a declining population," *American Economic Review* 112(11): 3489–3527.

Kallis, G., et al. (2018) "Research on degrowth," *Annual Review of Environment and Resources* 43: 291–316.

Kerschner, C., et al. (2018) "Degrowth and technology: Towards feasible, viable, appropriate and convivial imaginaries," *Journal of Cleaner Production* 197: 1619–36.

Kohler, H., Behrman, J. R., and Skytthe, A. (2005) "Partner + children = happiness? The effects of partnerships and fertility on well-being," *Population and development review* 31(3): 407–45.

Kopnina, H. and Washington, H. (2016) "Discussing Why Population Growth Is Still Ignored or Denied," *Chinese Journal of Population Resources and Environment* 14(2): 133–43. https://doi.org/10.1080/10042857.2016.1149296.

Latouche, S. ([2007] 2009) *Farewell to Growth*. Polity Press.

Lorek, S. and Fuchs, D. (2013) "Strong sustainable consumption governance – precondition for a degrowth path?" *Journal of Cleaner Production* 38: 36–43. https://doi.org/10.1016/j.jclepro.2011.08.008.

Miller, J. C. and Stovall, T. (2019) "The 'right to consume'? Re-thinking the dynamics of exclusion/inclusion in consumer society," *Consumption Markets & Culture* 22(5–6): 568–81. https://doi.org/10.1080/10253866.2018.1562712.

Mouton, F., et al. (2015) "Necessity for ethics in social engineering research," *Computers & Security*, 55, 114–127. https://doi.org/10.1016/j.cose.2015.09.001

Nelson-Coffey, S. K. (2018) "Married ... with Children: The Science of Well-Being in Marriage and Family Life," in Diener, Oishi, Tay (2018) *Handbook of Well-Being*. DEF Publishers.

Newman, P. A., et al. (2009) "What would have happened to the ozone layer if chlorofluorocarbons (CFCs) had not been regulated?" *Atmospheric Chemistry and Physics* 9(6): 2113–2128. https://doi.org/10.5194/acp-9-2113-2009

Nyyssönen, J. (2013) "Sami counter-narratives of colonial Finland. Articulation, reception and the boundaries of the politically possible," *Acta Borealia* 30(1): 101–121. https://doi.org/10.1080/08003831.2013.776738.

Oishi, S. and Diener, E. (2014) "Residents of Poor Nations Have a Greater Sense of Meaning in Life Than Residents of Wealthy Nations," *Psychological Science* 25(2): 422–30. https://doi.org/10.1177/0956797613507286.

Parrique, T., et al. (2019) *Decoupling debunked: Evidence and arguments against green growth as a sole strategy for sustainability.* European Environmental Bureau. https://eeb.org/wp-content/uploads/2019/07/Decoupling-Debunked.pdf.

Raworth, K. (2012) "A Safe and Just Space for Humanity: Can We Live within the Doughnut," *Oxfam Discussion Paper*. https://doi.org/10.1163/2210-7975_HRD-9824-0069

Roberts, D. (2018) "I'm an environmental journalist, but I never write about overpopulation. Here's why." *Vox*, 29 November 2018. https://www.vox.com/energy-and-environment/2017/9/26/16356524/the-population-question.

Ruuska, T., Heikkurinen, P., Levasseur, T., and Gare, A. (2024) Redefining Violence for the Anthropocene: From Ecocide to Ecological Civilization. *Cosmos and History: The Journal of Natural and Social Philosophy*, 20(2), 56–76.

Ruuska, T. (2017) *Reproduction of Capitalism in the 21st Century: Higher Education and Ecological Crisis*. Ph.D. diss., Aalto University.

Sen, A. (1999) *Development as Freedom*. Knopf.

Thiel, P. L., and Hallgren, H. (2018) "Rights of Nature as a Prerequisite for Sustainability," in *Strongly Sustainable Societies*, edited by Bonnedahl and Heikkurinen. Routledge.

Ulvila, M. and Pasanen, J. (2009) *Sustainable Futures: Replacing Growth Imperative and Hierarchies with Sustainable Ways*. Ministry of Foreign Affairs of Finland.

Ulvila, M. and Wilén, K. (2017) "Engaging with the Plutocene," in *Sustainability and Peaceful Coexistence for the Anthropocene*, edited by P. Heikkurinen. Routledge, 119–39. http://dx.doi.org/10.4324/9781315206820-9.

UN (1999). *The World at Six Billion*. Population Division, Department of Economic and Social Affairs. United Nations.

UN (2015) *World Population Prospects: The 2015 Revision* (DVD ed.). File POP/1–1: Total Population (Both Sexes Combined) by Major Area, Region and Country, Annually for 1950–20100 (Thousands). United Nations, Department of Economic and Social Affairs, Population Division. Retrieved April 22, 2017, from http://esa.un.org/unpd/wpp/Download/Standard/Population

Victor, P. A. (2008) *Managing Without Growth: Slower by design, not disaster.* Edward Elgar Publishing.

Wackernagel, M. and Rees, W. (1998) *Our Ecological Footprint: Reducing human impact on the earth* (*Vol. 9*). New Society Publishers.

Wake, D. B. and Vredenburg, V. T. (2008) "Colloquium Paper: Are We in the Midst of the Sixth Mass Extinction? A View from the World of Amphibians," *Proceedings of the National Academy of Sciences of the United States of America* 105 Suppl 1: 11466–73. https://doi.org/10.1073/pnas.0801921105.

Wiedmann, T. O., et al. (2015) "The Material Footprint of Nations," *Proceedings of the National Academy of Sciences of the United States of America* 112(20): 6271–76. https://doi.org/10.1073/pnas.1220362110.

Wynes, S. and Nicholas, K. A. (2017) "The climate mitigation gap: education and government recommendations miss the most effective individual actions," *Environmental Research Letters* 12(7): 074024. https://doi.org/10.1088/1748-9326/aa7541.

CHAPTER 12

Too Many People: Applying IPAT in an Overpopulated World

Philip Cafaro

1 Introduction

The IPAT equation was originally formulated as a framework for understanding the fundamental causes of environmental problems and guiding effective policy solutions. The success of the Kaya identity (described below) in helping climate scientists explain and predict changes in global CO_2 emissions has vindicated IPAT as a tool for furthering understanding. This chapter uses the IPAT framework to explore causes and possible solutions to the two defining environmental problems of our age: climate change and mass species extinction (sections 3 and 4). Finding strong evidence that these problems are unsolvable in a world with 8 billion people, it attempts to specify just and effective policies to substantially reduce the current human population—while remembering that environmental policy must address all three IPAT factors to have any chance of success (section 5). I begin by exploring why discussing population is so hard.

2 Initial Considerations

2.1 *Some History of Population Debates*

Contemporary environmental advocates generally shun population issues, but this was not always so. In the 1960s and 1970s, most people saw population stabilization as a key part of a complete environmentalism. Many founders of the global environmental movement were biologists who focused on competition for finite resources and spoke of the carrying capacity of land. It was natural for them to think of people as potentially pushing beyond nature's human carrying capacity and thus causing environmental problems. In addition, many of these scientists-turned-advocates were intimately acquainted with wild nature, leading them to question the justice of human over-appropriation of the habitat and resources other species needed for survival. Aldo Leopold, for example, wrote in *A Sand County Almanac* that a true land ethic must affirm

 | DOI:10.1163/9789004746404_013

other species' "right to continued existence, and, at least in spots, their continued existence in a natural state."[1]

The best-selling U.S. environmental tract of the 1960s was *The Population Bomb* by Paul Ehrlich. In his view, humanity's rapidly expanding numbers were at the root of our ever-increasing demands on the biosphere. Ehrlich and John Holdren popularized a standard formula for considering human beings' environmental impacts: I = P × A × T, where P equals population, A equals per capita consumption or "affluence," and T represents how efficiently current technologies service human demands.[2] Increasing a nation's numbers or its affluence will increase the human impact on the environment; increasing both intensifies that impact. Technological changes can help limit environmental impacts, in particular cases. But T cannot drive environmental impacts (I) down to zero and the overall trend across human history has been for technological innovations to increase resource use and pollution. Ehrlich emphasized that technology cannot be our environmental savior, particularly since technological changes are not typically driven by sustainability concerns, but rather by the profit motive.

When the first Earth Day was held in 1970, the global population was 3.7 billion—less than half what it is today—yet many speakers around the world discussed the need for population stabilization. In the late 1960s, Ehrlich and other environmentalists founded Zero Population Growth (ZPG) to work for population stabilization in the United States and around the world. As Roy Beck and Leon Kolankiewicz note in a study of the rise and fall of population concern in the United States, "in the 1970s, ZPG's population policy recommendations covered every contribution to U.S. population growth. It included stands on contraceptives, sex education for teenagers, equality for women, abortion, opposition to illegal immigration, and proposals to reduce legal immigration from about 400,000 a year to 150,000 a year by 1985 in order to reach zero population growth by 2008."[3] Other environmental groups took similarly strong stands in favor of population stabilization. In 1969, for example, the board of directors of the Sierra Club "urged the people of the United States to abandon population growth as a pattern and goal; to commit themselves to limit the total population of the United States in order to achieve a

1 A. Leopold, *A Sand County Almanac with Essays on Conservation from Round River* (Oxford University Press, 1966).

2 P. Ehrlich and J. Holdren, "One-dimensional ecology," *Bulletin of the Atomic Scientists* (1972).

3 R. Beck and L. Kolankiewicz, "The environmental movement's retreat from advocating U.S. population stabilization (1970–1998): A first draft of history," *Journal of Policy History* 12 (2000).

balance between population and resources; and to achieve a stable population no later than the year 1990."

How times have changed! Today environmental groups in the developed world are largely silent regarding population growth. They rarely weigh in on national population policies, leaving that to corporate lobbyists and pro-growth economists. Key reasons for this retreat in the U.S. included declining fertility (leading many to conclude that population growth had ended) and the heating up of anti-abortion politics. Also important in the retreat from population activism was the fact that immigration replaced native fertility as the main driver of population growth throughout the developed world. In fact, racial concerns made developed-world environmentalists nervous about addressing the main drivers of continued population growth in both richer countries (immigration) and poorer ones (high fertility rates). Left unaddressed was whether environmental goals could be achieved in a context of unprecedented human numbers and continued population growth.

We now know they can't. Humanity is cooking the Earth, bathing it in a toxic brew of plastics and chemicals, and extinguishing many of its species. None of these global environmental trends are improving; instead, they are merging into a single calamity imperiling the entire living world, including us. During the past decade, many scientists have concluded that addressing population must be part of global industrial civilization reforming itself sufficiently to avoid ecological disaster. A recent "Warning of a Climate Emergency," signed by over 11,000 scientists, forthrightly describes continued increases in human population and the world's GDP as "profoundly troubling signs" of ecological decline and states: "the world population must be stabilized—and, ideally, gradually reduced—within a framework that ensures social integrity."[4] It turns out a lot rides on talking honestly about population matters.

2.2 *Population or Overpopulation?*

According to David Attenborough, "all of our environmental problems become easier to solve with fewer people, and harder—and ultimately impossible—to solve with ever more people."[5] Attenborough has seen the global population triple since he narrated his first TV nature documentary in the 1950s, as people have, in his words, "overrun the world." So he advocates for policies to end population growth and stabilize global population sooner rather than later: above all, universal access to effective modern contraception.

4 W. Ripple et al., "World Scientists' warning of a climate emergency," *Bioscience* (2020).

5 D. Attenborough, "Impact of population growth on the planet," lecture to the Royal Society for the Encouragement of Arts, Manufactures and Commerce (London, 2011).

Sir David avoids the term 'overpopulation,' as do most environmentalists today. Perhaps it isn't necessary to show that our societies, or the whole world, are overpopulated. If we can convince people that addressing human numbers would help solve environmental problems, that might be enough to justify good population policies.

Consider global climate change. The Intergovernmental Panel on Climate Change assures us that population growth has been an important factor in driving increases in carbon emissions during the past four decades, accounting for about 40% of the global increase, with the other 60% driven by increased affluence (see section 3). On that basis Brian Walsh, Detlef van Vuuren, Brian O'Neill, and others, have advocated for policies to reduce future population growth—without saying the world is overpopulated, or venturing into the complexities of what a sustainable number of human beings might actually be.[6]

Arguably population growth has been similarly important in driving recent biodiversity losses, ocean acidification, and the whole array of global environmental problems (see Figure 12.1).[7] The world's nations have failed to address these problems successfully and the dangers of continued failure are immense. That can justify policies to limit population growth—all without using the word 'overpopulation.' According to the IPAT equation, human numbers, human affluence, and the technological means we use to accommodate them are of equal and fundamental importance in driving environmental impacts.[8] We can show population's importance without explicitly stating that any current or future population is too high.

A potential problem with this approach is that it doesn't *prove* that it is necessary to address population. Perhaps technological fixes, voluntary consumption reductions, clever efficiency improvements, or some combination would be sufficient, or better for some reason. Population advocates can respond that humanity's failure to successfully address global environmental problems in recent decades shows the need to tackle all three factors simultaneously. As Michel Bourban argues eloquently in this volume, the time is past

6 B. Walsh et al., "Pathways for balancing CO_2 emissions and sinks," *Nature Communications* 8 (2017); D. van Vuuren et al., "Alternative pathways to the 1.5 C target reduce the need for negative emission technologies," *Nature Climate Change* 8 (2018); B. O'Neill et al., "Demographic change and carbon dioxide emissions," *The Lancet* 380 (2012).

7 International Union for the Conservation of Nature (IUCN), "Importance for the conservation of nature of removing barriers to rights-based voluntary family planning," motion passed at IUCN World Conservation Congress (Marseille, 2020); K. Higgs, "Limits to growth: human economy and planetary boundaries," *Journal of Population and Sustainability* 2 (2017).

8 It should be noted, however, that Ehrlich and Holdren introduced the IPAT equation to stress the importance of P, which they believed was relatively neglected compared to A and T.

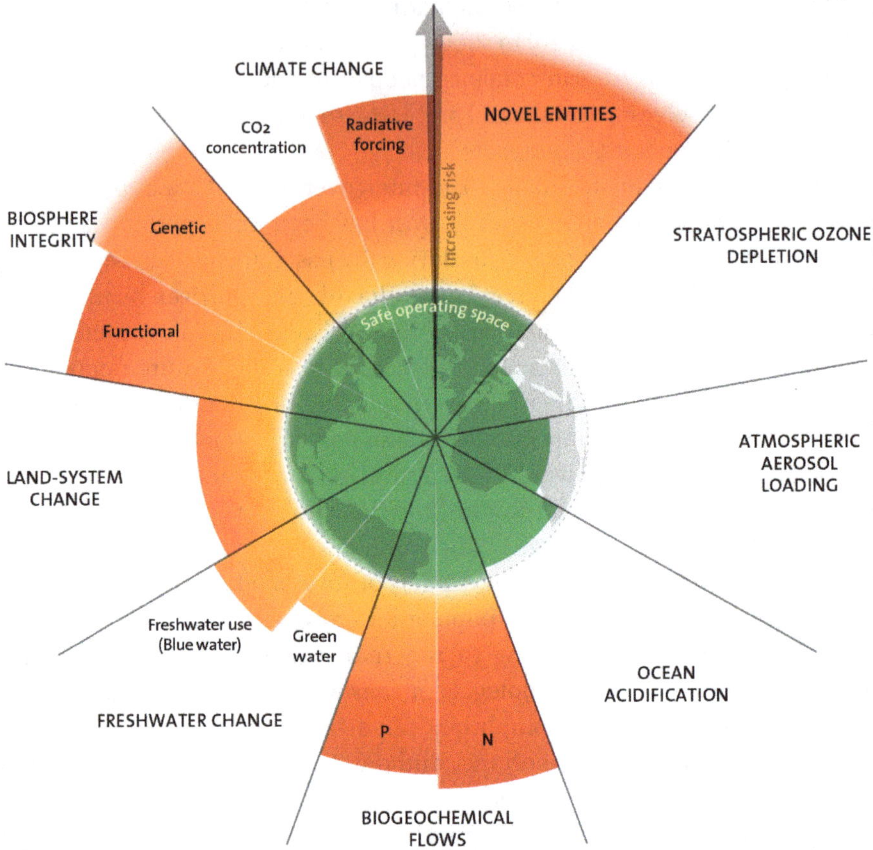

FIGURE 12.1 The nine planetary boundaries (2023 update). See also the 2025 update: https://www.stockholmresilience.org/news--events/general-news/2025-09-24-seven-of-nine-planetary-boundaries-now-breached.html, "Seven of nine planetary boundaries now breached" (Planetary Health Check 2025).

for speculating about ideal solutions, or misusing IPAT to prove the pre-eminence of one or another factor. Instead, we need strong efforts on all three fronts, given that continued failure is just too dangerous. This fits the views of most population activists, who rarely argue for limiting population instead of decreasing per capita consumption or deploying clever technologies, but for a comprehensive approach. A convincing defense of addressing population will also depend on specifying policies to end or reverse population growth that do not violate human rights. Of course, the same holds true for proposals to reduce per capita consumption or deploy new technologies.

2.3 *Overpopulation a Necessary Term*

One can advocate for population action without mentioning overpopulation. Nevertheless, it appears to be a necessary term, both to capture humanity's current situation and to point the way toward a sustainable future.[9] I define overpopulation as follows:

> Human societies, or the world as a whole, are overpopulated when their populations are too large to preserve necessary natural resources and ecosystem services for future human generations, or to share those resources and services justly with other species.

Recent estimates of a sustainable global population tend to run between 1.5 and 4 billion people, depending on how optimistic researchers are about voluntary consumption reductions, future technologies, and international cooperation. These estimates all affirm that the more people we want to sustain, the more modest their average standard of living must be. They assume increased efficiency and a widespread willingness to limit consumption—not *instead* of addressing population, but *in order to* sustain one quarter to one half of Earth's current population.[10]

The costs of failing to create sustainable societies will be continued environmental degradation and loss of the ecosystem services our descendants will need to live good lives; perhaps even a "ghastly future."[11] So there is a clear moral imperative to create sustainable societies. But we are a long way from sustainability and these studies suggest it will be difficult to achieve, even with much lower populations. Hence we need the term 'overpopulation' to explain that the current global population is probably much too large to ever be sustainable and to advocate for policies to significantly reduce bloated *national* populations, in both rich and poor countries, so future citizens can live flourishing lives while not using more than their fair share of limited resources.

The second half of my definition states societies are overpopulated if their populations are too large to share Earth justly with other species. This assumes

9 F. Götmark et al., "Discussing Population Concepts: Overpopulation is a Necessary Word and an Inconvenient Truth," *Indian Journal of Population and Development* 1 (2021).

10 P. Dasgupta, *Time and the Generations: Population Ethics for a Diminishing Planet* (Columbia University Press, 2019); T. Lianos and A. Pseiridis, "Sustainable welfare and optimum population size," *Environmental Development and Sustainability* 18 (2016): 1679–1699; C. Tucker, *A Planet of 3 Billion: Mapping Humanity's Long History of Ecological Destruction and Finding Our Way to a Resilient Future* (Observatory Press, 2019).

11 C. Bradshaw et al., "Underestimating the challenges of avoiding a ghastly future," *Frontiers in Conservation Science* 1 (2021).

that other species have a right to continued existence free from untimely anthropogenic extinction. Most discussions of Earth's human carrying capacity assume people have a right to monopolize the world's resources: that other species are themselves nothing more than resources for our use. If environmentalists ignore the P factor, we accept this selfish view and allow it to govern our interactions with other species. That's because more people equal fewer wild lands and less wildlife.[12]

In just the past fifty years, wild vertebrate populations have declined by approximately 70% globally. Anthropogenic extinction levels are an estimated 1000 times higher than the historical background rate and climbing. Such trends provide *prima facie* evidence that humanity is overpopulated. The key to reversing them is preserving much more wildlife habitat in protected areas (see section 4). But such interspecies justice is impossible at anything like our current population. To grant our fellow earthlings sufficient space to avoid a mass extinction, we will need billions fewer people. Instead, we are on track for billions more.[13]

Sharing Earth fairly with other species and with future human generations are noble goals. We can achieve them, but only with much smaller human populations. The following two sections detail the evidence for this regarding climate change and biodiversity loss.

3 Climate Change and Population

3.1 *IPAT Affirmed by the Scientists*

Working group III's contribution to the Intergovernmental Panel on Climate Change's *Sixth Assessment Report on Climate Change* came out in summer of 2022, summarizing the recent scientific literature on the causes of climate

12 W. Staples III and P. Cafaro, "For a Species Right to Exist," in P. Cafaro and E. Crist (eds.), *Life on the Brink: Environmentalists Confront Overpopulation* (University of Georgia Press, 2012); P. Cafaro, "Reducing human numbers and the size of our economies is necessary to avoid a mass extinction and share Earth justly with other species," *Philosophia* 50 (2022).

13 World Wildlife Fund, *Living Planet Report 2022—Building a nature-positive society* (Gland, Switzerland); Stuart Pimm et al., "The biodiversity of species and their rates of extinction, distribution, and protection," *Science* 344 (2014); E. Crist et al., "Protecting half the planet and transforming human systems are complementary goals," *Frontiers of Conservation Science* 2 (2021); United Nations, *World Population Prospects 2022*, Population Division, Department of Economic and Social Affairs.

change and current emissions trends.[14] This mitigation report also assessed humanity's options to limit future climate change and the possible environmental impacts of different policies. An opinion piece in *The Guardian* titled "Scientists have just told us how to solve the climate crisis—will the world listen?" called the new report "a major leap forward compared to previous reports." According to author Simon Lewis, it broke new ground by rejecting population growth as a driver of increased greenhouse gas emissions. Lewis wrote:

> The last IPCC summary on solutions in 2014 labelled population growth as one of 'the most important drivers of increases in CO_2 emissions from fossil fuel combustion.' Such dangerous misunderstandings are now gone. Seven years on, these old 'blame the poor' arguments increasingly seem like a relic of a previous age.[15]

For decades, climate scientists have used the Kaya identity to explain changes in global CO_2 emissions: Emissions = Population × GDP per capita × Energy used per unit of GDP × CO_2 generated per unit of energy. This is just a version of IPAT with two variables used to capture the T factor. Based on this framework, past IPCC assessment reports identified demographic and economic growth as the main drivers of climate disruption. Had that really changed? I decided to look more closely at the IPCC's new report.

The full mitigation report, the one written by scientists, is 2,707 pages long. In it, the fundamental role of growth in driving higher greenhouse gas emissions is clear, as it has been in past reports. Chapter 2 of the full report, "Emissions Trends and Drivers," notes:

> Globally, GDP per capita and population growth remained the strongest drivers of CO_2 emissions from fossil fuel combustion in the last decade (robust evidence, high agreement). Trends since 1990 continued in the years 2010 to 2019 with GDP per capita and population growth increasing emissions by 2.3% and 1.2% yr-1, respectively. This growth outpaced the reduction in the use of energy per unit of GDP (−2% yr-1, globally) as well as improvements in the carbon intensity of energy (−0.3% yr-1).[16]

---

14 Intergovernmental Panel on Climate Change (IPCC), *Climate Change 2022: Mitigation of Climate Change. Contribution of Working Group III to the Sixth Assessment Report of the Intergovernmental Panel on Climate Change.*

15 S. Lewis, "Scientists have just told us how to solve the climate crisis—will the world listen?" *The Guardian* (April 6, 2022).

16 IPCC 2022, pp. 2–4.

This paragraph is reiterated word for word in the *Technical Summary* to the main report, which was also written by the scientists.

The *Technical Summary* also notes that demographic and economic growth are likely to continue driving higher emissions. It discusses this in relation to the so-called Shared Socioeconomic Pathways (SSPs) that model different policy scenarios going forward:

> The main emissions drivers across the SSPs include growth in population reaching 8.5–9.7 billion by 2050, and an increase in global GDP of 2.7–4.1% per year between 2015 and 2050. Final energy demand in the absence of any new climate policies is projected to grow ... The highest emissions scenarios in the literature result in global warming of >5°C by 2100, based on assumptions of rapid economic growth and pervasive climate policy failures.[17]

According to the scientists, the P and A factors are indeed the main drivers of increased greenhouse gas emissions. Population growth compounds the effects of increased wealth by the number of people enriched (and vice versa) and will be a major influence on emissions going forward. To note just one more place in chapter 1 where this point is made unambiguously:

> Global GHG emissions have continued to rise since AR5, though the average rate of emissions growth slowed, from 2.4% (from 2000–2010) to 1.3% for 2010–2019. ... Important driving factors include population and GDP growth. The pause in emissions growth reflected interplay of strong energy efficiency improvements and low-carbon technology deployment, but these did not expand fast enough to offset the continued pressures for overall growth at [the] global level.[18]

At numerous places in the report, the authors state that future population growth could undermine mitigation efforts in specific economic sectors, including heavy industry and agriculture. There is also a reminder of the ubiquity of population's impacts on emissions, including in the developed world:

> Population growth has also remained a strong and persistent upward driver in almost all regions (+ 1.2% yr-1 globally from 2010 to 2019), although per capita emission levels are very uneven across world regions. Therefore, modest population increases in wealthy countries may have a

17 Ibid., p. TS-41.

18 Ibid., pp. 1–13.

similar impact on emissions as high population increases in regions with low per capita emission levels.[19]

Population's importance is reinforced by a figure from the main report (reproduced here as Figure 12.2), which provides data for regional changes in the four Kaya factors from 2010 to 2019. Population and per capita affluence both grew in every region of the world, overwhelming technological efficiency improvements and driving continued global increases in carbon emissions.

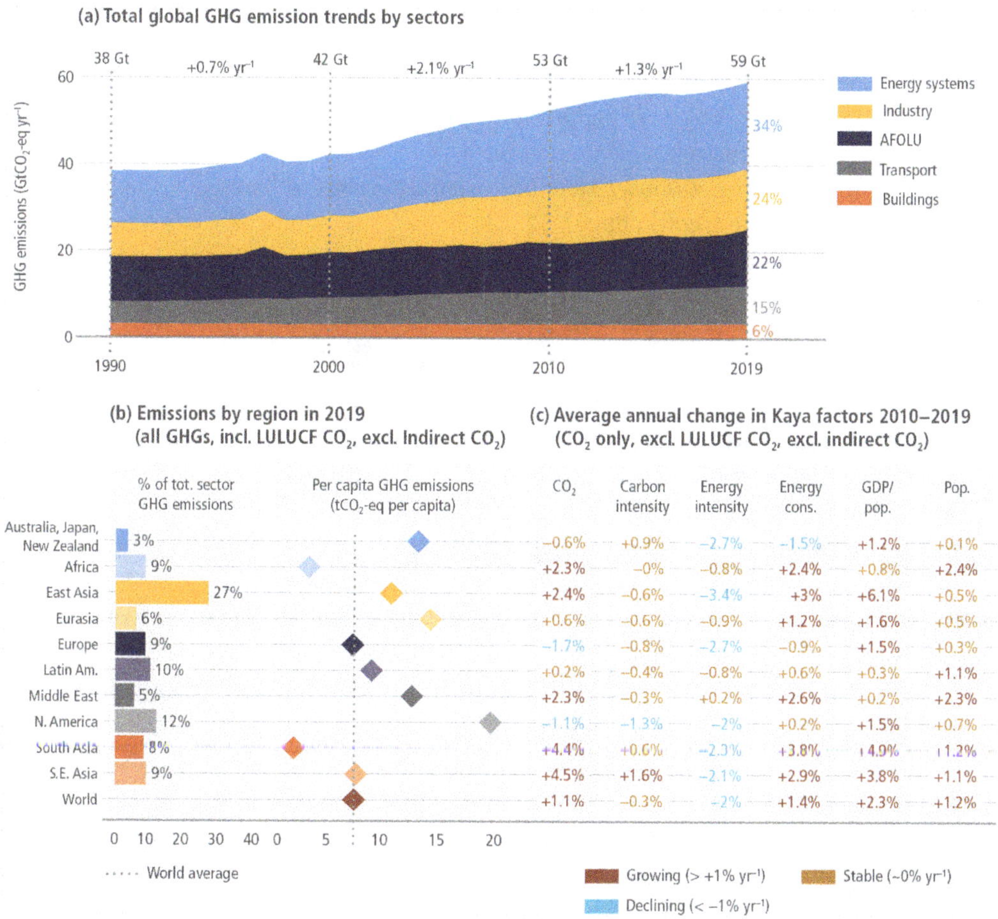

Region	% of tot. sector GHG emissions	CO2	Carbon intensity	Energy intensity	Energy cons.	GDP/pop.	Pop.
Australia, Japan, New Zealand	3%	−0.6%	+0.9%	−2.7%	−1.5%	+1.2%	+0.1%
Africa	9%	+2.3%	−0%	−0.8%	+2.4%	+0.8%	+2.4%
East Asia	27%	+2.4%	−0.6%	−3.4%	+3%	+6.1%	+0.5%
Eurasia	6%	+0.6%	−0.6%	−0.9%	+1.2%	+1.6%	+0.5%
Europe	9%	−1.7%	−0.8%	−2.7%	−0.9%	+1.5%	+0.3%
Latin Am.	10%	+0.2%	−0.4%	−0.8%	+0.6%	+0.3%	+1.1%
Middle East	5%	+2.3%	−0.3%	+0.2%	+2.6%	+0.2%	+2.3%
N. America	12%	−1.1%	−1.3%	−2%	+0.2%	+1.5%	+0.7%
South Asia	8%	+4.4%	+0.6%	−2.3%	+3.8%	+4.9%	+1.2%
S.E. Asia	9%	+4.5%	+1.6%	−2.1%	+2.9%	+3.8%	+1.1%
World		+1.1%	−0.3%	−2%	+1.4%	+2.3%	+1.2%

FIGURE 12.2 Trends and drivers of global GHG emissions, including b) share of total and per capita GHG emissions by world region in 2019, and c) Kaya decomposition of CO_2 emissions drivers
SOURCE: IPCC 2022 (FIGURE 2.16, IN PART)

19 Ibid., pp. 2–47, 2–53, 2–42.

3.2 *IPAT Censored by Politicians and the Media*

Given all this, how could Simon Lewis claim that the IPCC's new mitigation report avoided the "dangerous misunderstanding" that population growth is an important driver of increased greenhouse gas emissions? The answer is simple: none of this discussion of population and economic growth made it into the IPCC's short *Summary for Policymakers*, which is all most people read (including reporters). Unlike the full report, this document is vetted by political appointees. They apparently decided to censor anything that might call into question the goodness of continued growth.

For example, the chapter of the full report on emissions trends and drivers makes three bold-faced statements:

> Global net anthropogenic greenhouse gas emissions during the last decade (2010–2019) were higher than at any previous time in human history (high confidence).
>
> Emissions growth has varied, but persisted across all groups of greenhouse gases (high confidence).
>
> Globally, GDP per capita and population growth remained the strongest drivers of CO_2 emissions from fossil fuel combustion in the last decade (robust evidence, high agreement).[20]

The *Summary for Policymakers* repeats and elaborates on the first two sentences. But the third one is dropped; instead, its authors skip ahead to the more congenial topic of regional disparities in per capita greenhouse gas emissions.[21] We are thus left in the strange situation where the chief public-facing document from the IPCC's new mitigation report, the one most people saw, did not clearly identify the fundamental drivers of global climate change! This is so even though this key information is contained in the full report the summary is supposed to be summarizing.

As the *Summary for Policymakers* proceeds to consider policies to limit climate change, P and A, the two factors of the Kaya identity moving rapidly in the wrong direction, are simply excluded from consideration. So it is not surprising that the words 'contraception,' 'family planning,' 'overconsumption,' 'temperance,' and 'limits to growth' do not appear. Meanwhile, measures to address the T factor are discussed at length, not because they have been shown to be easier,

20 Ibid., pp. 2–4.

21 Ibid., pp. SPM 4–8.

cheaper, or more equitable, but by default.[22] The politicians who censored the summary are even willing to consider dangerous and unjust technological fixes such as geoengineering, rather than contemplate limits to growth.

Is this progress, as *The Guardian* claims? Or just the latest example of our inability to address global environmental decline honestly and with all the tools at our disposal?

3.3 *Aborting a Wider Debate about Growth*

Several other valuable discussions in the IPCC's full mitigation report seem to have left no trace on the *Summary for Policymakers*. Section 1.6 in chapter 1, "Achieving mitigation in the context of sustainable development," primarily discusses this in mainstream economic terms, with the possibility and goodness of increasing P and A unquestioned. But it does note the inherent connection between economic development and increased greenhouse emissions. More broadly, it questions whether mainstream economic approaches can guide the world to sustainability. For example:

> The 'green economy' and green growth—growth without undermining ecological systems, partly by gaining economic value from cleaner technologies and systems and that is inclusive and equitable in its outcomes—has gained popularity in both developed and developing countries as an approach for harnessing economic growth to address environmental issues. Critics however argue that [the] green economy ultimately emphasises economic growth to the detriment of other important aspects of human welfare such as social justice, and challenge the central idea that it is possible to decouple economic activity and growth (measured as GDP increment) from increasing use of biophysical resources (raw materials, energy). ... Literature on degrowth, post growth, and post development questions the sustainability and imperative of more growth especially in already industrialised countries and argues that prosperity and the 'Good Life' are not immutably tied to economic growth.[23]

22 P. Cafaro, "Avoiding catastrophic climate change: Why technological innovation is necessary but not sufficient," in Ronald Sandler (ed.), *Ethics and Emerging Technologies* (Palgrave-Macmillan, 2014).

23 IPCC 2022, pp. 1–41.

A similar discussion in chapter 1 of the full report, on climate ethics, questions the sufficiency of mainstream techno-managerial approaches to mitigation:

> A large body of literature examines the critical role of values, ethics, attitudes, and behaviours as foundational frames for understanding and assessing climate action, sustainable development and societal transformation. Most of this work is offered as a counterpoint or critique to mainstream literature's focus on safe-guarding of economic growth of nations, corporations and individuals. These perspectives highlight the dominance of economic utilitarianism in western philosophical thought as a key driver for unsustainable consumption and global environmental change. … While acknowledging the role of policy, technology, and finance, the 'managerialist' approaches that emphasise 'technical governance' and fail to challenge the deeper values that underpin societies may not secure the deep change required to avert dangerous climate change and other environmental challenges.[24]

Such discussions are new in an official IPCC document. They suggest changes to T may need to be supplemented by actions that limit P and A. But none of these fundamental economic and ethical debates made it into the *Summary for Policymakers*. There, instead, we find numerous cheery claims that technological innovations and managerial reforms can, by themselves, limit climate change to acceptable levels.

Of course, technological improvements have their place (T is also a fundamental factor in IPAT). But so does directly addressing the acknowledged drivers of climate change: excessive human numbers and overlarge human economies. The IPCC mitigation report fully supports doing so, with the scientists calling for a more robust discussion and a wider consideration of mitigation options.[25] But those who neutered the *Summary for Policymakers* apparently had other ideas. It is sad that *The Guardian* supported such censorship, rather than reporting on it. But it's not surprising, given the Left's general unwillingness to honestly discuss population issues. Ironically, this aversion nicely suits the imperatives of global capitalists seeking ever more workers and consumers. Unfortunately, it makes it more likely that climate policies will fail future generations.

24 Ibid., pp. 1–48.

25 See also Ripple et al., 2021.

4 Biodiversity Loss and Population

4.1 *Looming Mass Extinction*

While climate change has come to define environmentalism in the 21st century, biodiversity loss is an equally important global environmental problem. A scientific consensus exists that wild nature is rapidly dwindling. In 2010, the UN's Secretariat of the Convention on Biological Diversity estimated that humanity could extinguish one out of every three species on Earth within the next one to two hundred years, while according to Peter Raven and colleagues, "biodiversity is diminishing at a rate even faster than the last mass extinction at the end of the Cretaceous Period, 65 million years ago, with possibly two-thirds of existing terrestrial species likely to become extinct by the end of this century." Even using conservative estimates for current extinction rates and holding those rates steady, projecting them forward a few hundred years predicts immense losses.[26]

The United Nations has created a scientific panel modelled on the IPCC to summarize what is known about the causes, extent, and possible solutions to biodiversity loss: the Intergovernmental Panel on Biodiversity and Ecosystem Services (IPBES). Their first comprehensive *Global Assessment Report on Biodiversity and Ecosystem Services* summarizes human impacts on wild nature thus:

> Humanity is a dominant global influence on life on earth, and has caused natural terrestrial, freshwater and marine ecosystems to decline. Global indicators of ecosystem extent and condition have shown a decrease by an average of 47 percent of their estimated natural baselines, with many continuing to decline by at least 4 percent per decade.
>
> High-biodiversity tropical forests continue to dwindle, and global forest area is now approximately 68 percent of the estimated pre-industrial level. ... Only 13 percent of the wetlands present in 1700 remained by 2000; recent losses have been even more rapid.
>
> Marine ecosystems, from coastal to deep sea, now show the influence of human actions, with coastal marine ecosystems showing both large historical losses of extent and condition as well as rapid ongoing declines. ... Live coral cover on reefs has nearly halved in the past 150 years, the decline dramatically accelerating over the past two or three decades due to increased water temperature and ocean acidification.

26 G. Ceballos et al., "Accelerated modern human—induced species losses: entering the sixth mass extinction," *Science Advances* 1 (2015).

> Human actions have already driven at least 680 vertebrate species to extinction since 1500 … The proportion of species currently threatened with extinction according to the IUCN's Red List criteria averages around 25 percent … More than 40 percent of amphibian species, almost a third of reef-forming corals, sharks and shark relatives and over a third of marine mammals are currently threatened.[27]

In short, humanity is on the verge of causing a mass extinction of Earth's species. If these trends continue for another hundred years, there will be little wild nature left to worry about. A hundred years after that, our descendants could live on a completely humanized planet, crowded but strangely lonely. I believe preventing this mass extinction is a moral imperative, whether we consider the displacement of billions of individual non-human beings or the more fundamental and irreversible loss of millions of unique forms of life. It would be deeply wrong for humanity to monopolize a world capable of supporting such astounding diversity.[28]

4.2 *Biodiversity Loss Driven by Growth*

The cause of global biodiversity loss is clear: other species are being displaced by a rapidly growing human economy. *We* are replacing *them* with *us*, our economic support systems, our domestic animals, and our trash. From 1970 through 2020, the same period wild vertebrate populations declined 70%, human numbers doubled, the size of the global economy quadrupled, and international trade increased tenfold. The wildlife decline was caused by the human expansion. People took habitat and resources away from other species, because there were a lot more of us and our economy became more successful at transforming wild nature into resources for human use and profit—its primary goal. As the IPBES notes: "Today, humans extract more from the Earth and produce more waste than ever before."[29]

As is standard in the conservation biology literature, the IPBES report explains the causes of biodiversity loss in terms of five main "direct drivers," stating: "globally, land-use change is the direct driver with the largest relative impact on terrestrial and freshwater ecosystems, while direct exploitation of

27 IPBES, 2019, p. 24.

28 H. Rolston III, *Conserving Natural Value* (Columbia University Press, 1994); S. Donaldson and W. Kymlicka, *Zoopolis: A Political Theory of Animal Rights* (Oxford University Press, 2011).

29 S. Diaz et al., "Pervasive human-driven decline of life on earth points to the need for transformative change," *Science* 366 (2019): 1327; IPBES, 2019.

fish and seafood has the largest relative impact in the oceans. Climate change, pollution and invasive alien species have had a lower relative impact to date but are accelerating." The leading cause of terrestrial habitat loss is agricultural expansion: "over one third of the world's land surface and nearly three-quarters of available freshwater resources are devoted to crop or livestock production." In other words, as we feed more people more richly, less food and other resources are left for other species.[30]

According to the IPBES, human population increase "has strong implications for the degradation of nature," with "unsustainable use of the Earth's resources underpinned by a set of demographic and economic indirect drivers that have increased, and that furthermore interact in complex ways." The term 'indirect driver' is misleading—'fundamental causes' would be more accurate—yet the message is clear enough. "Anthropogenic drivers of biodiversity loss, including habitat loss as a result of land-use and sea use change, unsustainable agriculture, aquaculture and forestry, unsustainable fishing, pollution, and invasive alien species are [all] increasing globally." Growing human numbers and affluence are also the fundamental causes of humanity pushing beyond other planetary boundaries for safe use of the biosphere, such as ocean acidification, excess nitrogen deposition and, as we have seen, climate change (see again Figure 12.1).[31]

Just as population increases clearly contribute to biodiversity losses, so population *decreases* aid in restoring biodiversity. One sees this particularly clearly in Europe, which is densely populated, but also the first continent to end humanity's modern population explosion. Europe's overall population has stabilized in recent years and its rural population has declined 20% since 1960, contributing to extensive abandonment of less productive farmland. These trends have been valuable for wildlife, particularly larger herbivores and carnivores, which have naturally recolonized many former agricultural areas. Ecological restoration can accelerate these trends, turning what is often viewed as a negative (rural depopulation) into a positive (rewilding the landscape and restoring a country's natural heritage). From the Velebit Mountains in Croatia to Portugal's Côa Valley, from the Oder delta on the Baltic Sea to the Danube delta on the Black Sea, rewilding efforts have been facilitated by decreasing human populations. Similar examples can be found throughout the world.[32]

30 IPBES, 2019, p. 28; Crist et al., 2017.

31 IPBES, 2019, pp. 30, 33; Higgs, 2017.

32 H. Pereira and L. Navarro (eds.), *Rewilding European Landscapes* (Springer International, 2015); The Overpopulation Project, "Rewilding Success Stories" (2020).

4.3 *Protecting Habitat and Reducing Human Numbers*

Existing growth-focused economies are leading human societies deeper into ecological overshoot and creating a depauperate world. We need to transition to a new economic paradigm: one with economies designed to provide sufficient material goods (limiting A) for reasonable numbers of people (limiting P), not ever more stuff for ever more people forevermore.[33] To preserve global biodiversity and essential ecosystem services in the future, we must achieve absolute reductions in human resource use and pollution. We need to grow less food, drive fewer cars, fly fewer airplanes, pour less concrete. Obviously, having fewer people will facilitate all these goals.

It is true that increased efficiency and new technologies (improvements in T) may help decrease the environmental impacts of our economic activities, provided they are actually put to that use. A few sectors of the economy might benefit the environment if they grew: solar panel sales, for example. But given current realities, humanity must increase material and energy efficiencies through technological improvements *and* decrease the overall size of our economies. Our new task is to deploy clever technologies to shrink overall economic demands on the biosphere. There is little reason to think this can be accomplished by focusing exclusively on T and ignoring P and A. Nor is there good reason to focus on P to the exclusion of A or vice versa, in a world where decreasing either factor will involve real sacrifices. Because the task is urgent yet difficult, we must address all three factors now.

Protecting or restoring sufficient habitat is key to preserving other species. Conservation biologists calculate that setting half the globe's terrestrial and aquatic habitat off limits to intensive human economic uses could preserve 85–90% of species long-term; a higher percentage could be protected through efforts to safeguard particularly rich ecosystems. An international movement has sprung up committed to doing just that: Half Earth, or Nature Needs Half.[34] While their proposal is radical (today only about 15% of the Earth's land surface and 5.5% of its oceans are protected) it appears no less drastic limitations on human economic activity can keep us from exterminating a large percentage of the world's species. Half Earth has begun to enter the political mainstream: the Biden administration outlined plans to protect 30% of U.S. lands and

33 H. Daly and J. Farley, *Ecological Economics: Principles and Applications*, 2nd edition (Island Press, 2010); W. Rees, "Ecological economics for humanity's plague phase," *Ecological Economics* 169 (2020).

34 H. Locke, "Nature needs (at least) half: A necessary new agenda for protected areas," in G. Wuerthner et al. (eds.), *Protecting the Wild: Parks and Wilderness, the Foundation for Conservation* (Island Press, 2015); E. Dinerstein et al., "An ecoregion-based approach to protecting half the terrestrial realm," *Bioscience* 67 (2017).

ocean territories by 2030, while the European Parliament recently passed a resolution to protect 50% of the European Union's ecosystems by 2050.

To set aside enough habitat and sufficient resources to preserve robust populations of native species, most nations will have to significantly shrink their human populations.[35] Simply stabilizing already bloated populations will not be sufficient. Accepting a world of 8 billion people means accepting mass species extinction. So does accepting a global economy predicated on ever-increasing affluence. The reality is that overpopulation, high per capita consumption, and the excessive economic activity to which they inevitably lead are the fundamental causes of biodiversity loss. Smaller economies, made possible by reduced human numbers and more modest per capita consumption, are necessary to preserve what is left.

As with the P and A factors within human societies, there is a necessary trade-off between the amount of habitat and resources allocated to people and the amount available to the rest of life. More of one means less of the other. To be clear about my own position, I believe the goal of limiting human populations should not be to facilitate continued excessive per capita consumption in wealthy nations, or to duplicate this excess in poorer ones. Instead, we should create sustainable societies that provide a sufficiency for people, now and in the future, while sharing Earth justly with our non-human relations.[36]

5 Addressing Population

5.1 *How Overpopulated Are We?*

Like other sustainability advocates promoting reforms that run counter to mainstream economic boosterism, population advocates have learned to carefully police our language and policy proposals. Given how far societies are from sustainability, just slowing down their enthusiastic charge in the wrong direction can look like a reasonable goal. 'Population stabilization' sounds less threatening than 'population reduction.' But as Karin Kuhlemann observes, "that a population's size is stable in no way entails sustainability. It may be sustainable, or it may be far too large."[37]

That's the problem. Globally, current human numbers appear nowhere near compatible with long-term human wellbeing or the flourishing of other species.

35 Crist et al., 2021.

36 Cafaro, 2022.

37 K. Kuhlemann, "'Any size population will do?' the fallacy of aiming for stabilization of human numbers," *The Ecological Citizen* 1 (2018).

As previously noted, recent estimates of a sustainable global population range between 1.5 to 4 billion, even assuming relatively modest levels of per capita consumption. Most individual countries are also overpopulated, based on various plausible criteria: taking only their fair share of limited global resources; not degrading the environments within their national boundaries; their ability to sustain current human population levels long term. Acknowledging that all such estimates are simplifications that rely on controversial premises, let's look at one effort to specify sustainable national populations that takes P, A and T seriously, instead of emphasizing one factor at the expense of the others.

In their article "Sustainable welfare and optimum population size," Theodore Lianos and Anastasia Pseiridis calculate that the world could safely accommodate 3.1 billion people (P) living on an average annual income of $9,000 per person (A), an amount they deem sufficient for a satisfactory life at current technology levels (T).[38] They focus solely on Earth's human carrying capacity, relying on global footprint analysis to determine sustainable environmental impacts; leaving enough habitat and resources to avoid a mass species extinction would likely drive the sustainable human population number much lower. Lianos and Pseiridis then calculate sustainable populations for the world's 52 most populous nations, on the premise that each country is entitled to a share of the sustainable global population equal to its share of global cropland. Their approach does not presuppose that each country would only rely on the resources within its own borders; trade in agricultural and other products could still occur.

Other alternatives to fairly divide Earth's limited capacity to support people are conceivable. We might stipulate retaining individual nations' current percentage of the global population and shrinking all by the same percentage; we might base a fair division on nations' share of global economic output, or on the size of national territories. Any criterion will be controversial, but that should not prevent us from trying to specify a fair one. The key point is that whatever the criterion, allowing more people in one country means requiring fewer in another, if we seek to reduce overall numbers. Figure 12.3 shows the difference between current, sustainable, and projected populations for the world's five most populous countries based on Lianos and Pseiridis' stipulations and UN (2022) estimates.

38 Lianos and Pseiridis, 2016. Elizabeth Cripps (2021) asserts that an annual income of $9,000 is "comparatively affluent," but provides no examples of societies whose members are satisfied with a smaller one. While such calculations may seem arbitrary, they have the virtue of reminding readers of the trade-off between more people and more per capita wealth, if we are trying to limit overall demands on the environment. The more of one, the less we can have of the other, if IPAT is valid.

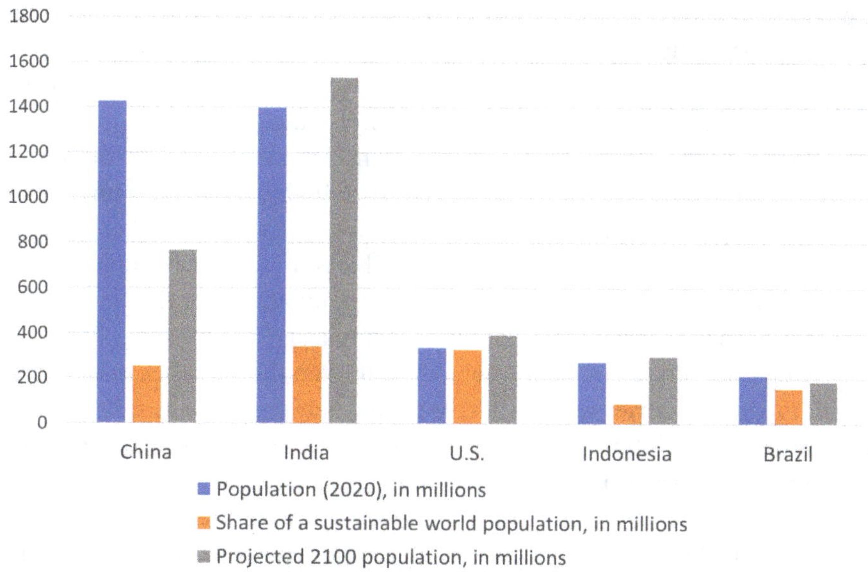

FIGURE 12.3 Population and overpopulation in the world's five most populous countries, 2020 and 2100
SOURCES: DATA IN FIRST AND LAST COLUMNS FROM UNITED NATIONS 2022, MIDDLE COLUMN FROM LIANOS AND PSEIRIDIS (2016)

Note that these sustainable national population estimates assume a willingness to limit or reduce average annual incomes to $9,000; at higher average incomes the sustainable population decreases proportionally. Of course, income is only a rough measure of natural resource use and waste generation. Many other things besides the availability of agricultural land factor into sustainability. Still, these calculations give some idea of the immense population decreases these five countries would need to achieve national sustainability and do their part for global sustainability.

China would have to cut its population by more than 1.15 billion people to reach sustainability: from over 1.4 billion today to approximately 250 million. India would also need to reduce its population by over a billion, down from 1.4 billion to about 340 million. At first glance, America looks in better shape, but only if we ignore the need to drastically cut per capita income and consumption to support 330 million people sustainably. Americans' average annual household income in 2020 was around six times greater than $9,000, so unless we were willing to cut our incomes substantially, a sustainable U.S. population would be a small fraction of our current numbers: 55 million at current levels of affluence.[39] Brazil, too, at first glance looks relatively close to a sustainable

39 Some of this decrease could come from drastically cutting the incomes of the wealthy, opening up space for more Americans living on lower incomes. Still, most Americans

population. But this assumes deforesting much of Brazil's portion of the Amazon rainforest to support the maximum number of Brazilians. If your definition of sustainability includes preserving the world's largest reservoir of terrestrial biodiversity, a sustainable Brazilian population is much lower.

Looking at the UN's projected populations for 2100, none of these five countries are anywhere close to achieving a sustainable population under current demographic trends and economic policies. Nor are most other nations, whether underdeveloped or overdeveloped. This is shown, for example, by Lianos and Pseiridis' calculations for sustainable populations for the seven most populous European nations, illustrated in Figure 12.4.

Leaving aside Russia as a continent-sized outlier, the six other most populous European countries would all have to cut their populations substantially to achieve sustainability: France by 40 percent, Italy by 68 percent, Germany by 71 percent, and the United Kingdom by a whopping 82 percent—down from 67 million to 12 million people. All this, remember, with the stipulation of an average annual income of $9,000. With the extra income that most Europeans

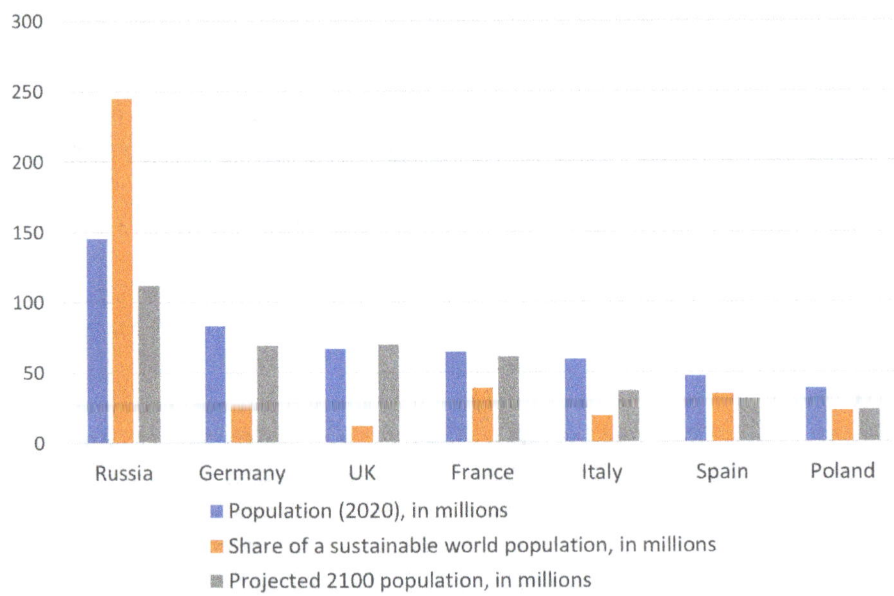

FIGURE 12.4 Population and overpopulation in Europe's seven most populous countries, 2020 and 2100.
SOURCES: DATA IN FIRST AND LAST COLUMNS FROM UNITED NATIONS 2022, MIDDLE COLUMN FROM LIANOS AND PSEIRIDIS (2016)

---

would need to accept greatly reduced incomes to sustain a population in the hundreds of millions.

probably would want to retain, sustainable population sizes decrease proportionally. They decrease even more as more habitat and resources are devoted to native biodiversity, which EU countries have formally committed to do.

Developing nations, meanwhile, face demands for more consumption by the poor, rising consumption by burgeoning middle classes, and the same excessive consumption by the wealthy seen in the developed world. Here, too, sustainability demands smaller human numbers, to free up ecological space for the poor to increase their consumption, the middle classes to enjoy newfound prosperity, and other species simply to survive.[40] Instead, many developing nations are moving rapidly in the opposite direction, as shown in calculations for sustainable populations for the six most populous sub-Saharan nations (Figure 12.5).

According to Lianos and Pseiridis' estimates, Nigeria would need to decrease its population of 208 million to 38% of its current size, 79 million, to achieve a sustainable population at an average annual income of $9,000. Instead, it

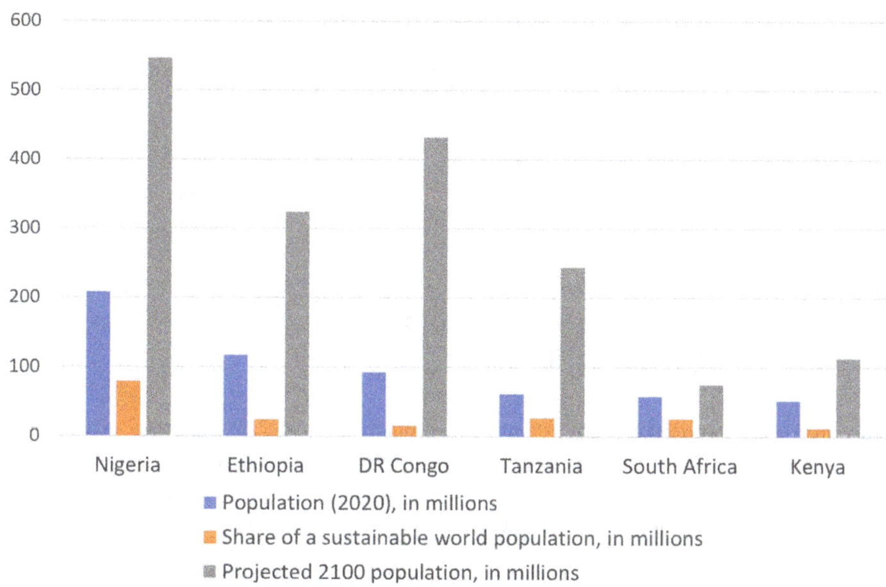

FIGURE 12.5 Population and overpopulation in sub-Saharan Africa's six most populous countries, 2020 and 2100

40 T. Hayward, "Ecological space: the concept and its ethical significance," in S. Gardiner and A. Thompson (eds.), *The Oxford Handbook of Environmental Ethics* (Oxford University Press, 2016).

is on track to increase more than 160% by 2100, to 546 million people. Ethiopia's current population of 117 million would need to shrink to 24 million, 20% of its current size, to achieve a sustainable population. Instead, it is currently projected to increase by 177% to 324 million. And so it goes throughout the region. Although these African nations currently have smaller per capita incomes and hence smaller per capita environmental impacts than "the Big 5" and most European nations, their citizens have no desire to remain poor. They want economic security and well-being.

The key point is that all societies, from the richest to the poorest, face the same trade-off between affluence and population when it comes to sustainability. The more of one, the less of the other—assuming a commitment to keeping total environmental impacts within sustainable bounds. Similarly, all societies must accept even smaller human numbers if they are committed to preserving or restoring robust populations of native biodiversity. And as we have seen, despite their differences in wealth, many developed and developing nations share a key characteristic: they are massively overpopulated relative to their actual or aspirational per capita consumption.

This conclusion holds whether we conceive sustainability as taking no more than our fair share of the global commons, or avoiding degrading environments within national boundaries. It holds regardless of how fairly we divide the sacrifices necessary to achieve sustainability within our societies. It even holds whether societies choose to preserve their native biodiversity or not. The UK and Nigeria might be able to squeeze a few million more people within their boundaries by sacrificing their remnant wildlife, the U.S. and Indonesia a few tens of millions more. But all four countries would still need to cut their populations drastically to have a chance to create sustainable societies and do their part to create a sustainable world.

5.2 *Just Population Policies*

If we want to create sustainable societies, we will need to decrease populations, reduce average consumption (at least in wealthy nations), and deploy less harmful technologies. All three, not one instead of the others. With our societies deep into ecological overshoot, environmentalists should push hard for action on all three fronts: P, A and T.[41]

41 E. Ganivet, "Growth in human population and consumption both need to be addressed to reach an ecologically sustainable future," *Environment, Development and Sustainability* 22 (2019); K. Shragg, *Move Upstream: A Call to Solve Overpopulation* (Freethought House, 2015).

I believe the potential consequences of humanity's current colossal overpopulation justify the following general ethical imperative:

> Would-be parents should restrict themselves to one biological child.

More is socially and environmentally irresponsible at this point in history. Many conscientious people around the world already limit themselves to two children, replacement rate, out of environmental concern. But as we have seen, merely stabilizing current populations will not be sufficient to avoid continued environmental degradation. Limiting procreation is the single most important contribution most individuals can make to sustainability. If populations are to decline to sustainable levels over the next one to two hundred years, many couples will have to embrace one-child families.[42]

As with other serious environmental goals, however, voluntarism alone will not achieve sustainable populations. Governments will have to enact policies explicitly designed to reduce future numbers of citizens. Which population policies nations choose and how necessary sacrifices are shared should be matters for democratic decision-making. Any policies enacted should respect human rights.[43]

With these caveats in mind, it seems to me the threat of global ecological catastrophe justifies the following three kinds of population policies, in both developed and developing nations:

> National governments should guarantee their citizens universal, free access to family planning services, modern contraception, and abortion on demand.

Having fewer children should be made as easy as possible. When women are free to choose whether to bear children and couples can limit the size of their families, fertility rates almost always decline. Such policies provide win/win/win solutions: upholding women's rights, decreasing poverty, and reducing

42 T. Meijers, "Climate change and the right to one child," in G. Bos and M. Düwell (eds.), *Human Rights and Sustainability* (Routledge, 2016); S. Wynes and K. Nicholas, "The climate mitigation gap: Education and government recommendations miss the most effective individual actions," *Environmental Research Letters* 12 (2017).

43 C. Hickey et al., "Population engineering and the fight against climate change," *Social Theory and Practice* 42 (2016); T. Hedberg, *The Environmental Impact of Overpopulation: The Ethics of Procreation* (Routledge, 2020).

human numbers.[44] But facilitating increased reproductive choice probably will not be sufficient to reduce populations quickly enough to avoid ecological disaster; hence the need for further policies that actively seek to reduce human numbers:

> National governments should encourage their citizens to have only one biological child and discourage them from having more: through information campaigns, direct advocacy, tax policies, safety net policies, and more.

Governments should educate their citizens on the many economic, environmental, and social benefits of small families—and incentivize them. We cannot rely solely on personal, voluntary choices to secure environmental sustainability. We cannot allow overly prolific free riders to overwhelm the efforts of more responsible citizens. It won't work, and our children and grandchildren will pay a harsh penalty when we fail.[45]And because *global* population decrease depends on strong *national* commitments to reducing national populations, a practical plan to reduce human numbers must also address immigration policies:

> National governments should limit and, in many cases, reduce immigration, as part of comprehensive efforts to reduce their populations to sustainable levels within three to four generations.

Currently many developed nations have growing populations despite fertility rates that have been below replacement rate for decades, due to mass immigration. We cannot ask citizens to limit their numbers for the common good, while their governments undermine their efforts by inviting in more people. We cannot ask people to consume less, while governments bring in more consumers. Again, it won't work. It isn't just, and common citizens will not accept it.[46]

Governments should not harshly punish people who have more than one child, or end all immigration. Instead, I propose population reduction as a major policy goal for countries that are overpopulated, with targeted policies to reduce their populations over time, including financial incentives for

---

44 K. Hardee et al., *Voluntary Family Planning Programs That Respect, Protect, and Fulfill Human Rights: A Conceptual Framework* (Futures Group, 2013).

45 A. Gheaus, "The right to parent and duties concerning future generations," *Journal of Political Philosophy* 24 (2016): 487–508.

46 P. Cafaro and J. O'Sullivan, "How Should Ecological Citizens Think About Immigration?" *Ecological Citizen* 3 (2019).

one-child families and financial penalties for families with more than two children. For example, rather than providing the same tax break for each additional child, as currently in the U.S., governments could provide a tax break for a first child, none for a second child, and a tax charge for additional children.[47]

I confess to some unhappiness with these conclusions. My proposed policies would entail significant costs to many would-be parents and immigrants. If they did not, they would not achieve their purpose: reducing human numbers. I would prefer not to have governments tell people where to live or how many children to have. However, in the crowded world we have created, such impositions seem preferable to massive ecological degradation and all that implies. Failure to support such policies now means accepting great human suffering and a biologically impoverished world in the future.[48]

I support similarly intrusive policies to incentivize lower average consumption and prohibit excessive consumption. For example, I would limit individuals to one round-trip airplane flight a year and prohibit personal aircraft; heavily subsidize mass transit while increasing taxes on personal vehicle use; heavily tax large homes; and so on down the line for all important categories of consumption.[49] In these cases, too, it would give me no pleasure to see governments stick their noses into such personal decisions, or discourage consumption that people find enjoyable and that allows other people to make an honest living. But I don't see a less intrusive path to achieving ecological sustainability. Do you?

5.3 *Objections and Responses*

Many will find these policy proposals overly restrictive, or even unjust. A common counterargument is that they would violate human rights. People have a right to choose the size of their families. People have a right to live and work where they want, regardless of national boundaries. People have a right to fly or own airplanes, if they didn't break any laws securing the money to buy their tickets or their planes (and can pass their flight tests).[50]

But these proposed rights, like all rights, are claims on limited resources. If anything like the ecological perils described by the IPCC and IPBES are correct, a right to have more than one child cannot be universalized due to resource

47 T. Rieder, *Toward a Small Family Ethic* (Springer, 2016).

48 E. Cripps, "Climate change, population, and justice: Hard choices to avoid tragic choices," *Global Justice: Theory Practice Rhetoric* 8 (2015).

49 P. Cafaro, "Reducing Consumption to Avert Catastrophic Global Climate Change: The Case of Aviation," *Natural Science* 5 (2013).

50 A. Hendrixson et al., "Confronting populationism: Feminist challenges to population control in an era of climate change," *Gender, Place & Culture* 27 (2019): 307–15; J. Carens, *The Ethics of Immigration* (Oxford University Press, 2013).

constraints. Not today and not for the next few generations. This fact is enough to show that no such defensible moral right currently exists, whatever our laws or traditions may say and however we may feel. As Immanuel Kant convincingly argues, no one can legitimately claim a right that cannot be universalized within his or her society. As John Stuart Mill adds, no one has a legitimate right to act in ways that significantly harm others; in this case, by degrading the ecosystem services that future generations will depend upon for their basic well-being.[51]

If we want to create societies where a moral right and *effective* legal rights to have more than one child exist in the future, we will need to accept a period where citizens are discouraged from having more than one child. Societies that fail to reduce their numbers proactively and humanely are likely to create conditions where it won't be possible to raise any children safely. That is the path we appear to be on now. Population policies must consider responsibilities as well as rights.[52]

A related argument holds regarding immigration, which involves individuals staking claims to limited resources in a new country. In an overpopulated world, a right to free international movement threatens the right of future citizens to have any children safely. It may undermine the rights of future children to food, shelter, and the stable civil order on which all rights depend. Starvation and the breakdown of civil society are forms of coercion too, and arguably worse ones than restrictions on immigration or procreation. Just as free-riding citizens cannot be allowed to undermine national efforts to achieve sustainability, neither can free-riding non-citizens, or free-riding nations that persist in 19th century fertility levels in the 21st century. Once again, I believe similar arguments hold regarding affluence. National economic policies should discourage unnecessary and excessive consumption in a world in rapid ecological decline. Limiting A is just as important as limiting P.[53] It is particularly important among the wealthy, in both developed and developing nations.

51 I. Kant, *Groundwork of the Metaphysics of Morals* (numerous editions, originally 1785); J. S. Mill, *On Liberty* (numerous editions, originally 1859).

52 Population Institute, *Demographic Vulnerability: Where Population Growth Poses the Greatest Challenges* (Washington, D.C., 2015); P. Cafaro, "Just population policies for an overpopulated world," *Ecological Citizen* 5 (2021): epub-046–1.

53 P. Cafaro, *How Many Is Too Many? The Progressive Argument for Reducing Immigration into the United States* (University of Chicago Press, 2015); T. Wiedmann et al., "Scientists' warning on affluence," *Nature Communications* 11 (2020): 3107.

In response to proposals to restrict immigration, some argue that "population is a global, not a national problem," others that "national borders are arbitrary."[54] To the first group, I say that overpopulation is both a national and a global environmental problem. Limiting national populations is necessary to meet important domestic responsibilities, such as securing sufficient food for fellow citizens and preserving sufficient habitat for native wildlife.[55] It is also the most important contribution people can make to limiting global population numbers, just as protecting forests in their own country is the primary contribution they can make to halting global deforestation, and reducing their own greenhouse gas emissions is the main contribution they can make toward mitigating global climate change. Think globally, *don't* act locally, is bad environmental advice.

To the second group, I say national borders are essential to the existence of nations, which are a positive good, not a necessary or unnecessary evil.[56] Nations allow individuals to pursue political projects that further the common good. They allow us to promote our interests and secure our rights; for example, by redistributing wealth, limiting corporate power, and enforcing laws curbing air and water pollution. They allow us to safeguard other species, too, by creating national parks and wildlife refuges, passing laws protecting endangered species, and forcing the rich and powerful to respect these decisions. As national citizens, we have a responsibility and an opportunity to create just and sustainable societies, while pursuing peaceful coexistence with our neighbors. Like democracy, a global order built on discrete nation-states and robust national governments is the worst possible system—except for all the alternatives.

To be clear, my responses to these objections should not be construed as denying the need to uphold human rights in setting population policies. Forced abortions or taxes that push children into poverty are wrong. Governments must not violate human rights to achieve public policy goals, even very important ones. Properly understood and implemented, rights uphold human dignity and further human flourishing. But pushed to extremes, they undermine them.[57] Rights must be balanced against one another and paired with

54 E. Crist, "Decoupling the global population problem from immigration issues," *The Ecological Citizen* 2 (2019); Carens, 2013.

55 L. Kolankiewicz et al., *From Sea to Sprawling Sea: Quantifying the Loss of Open Space in America* (Numbers USA, 2022).

56 Y. Tamir, *Why Nationalism* (Princeton University Press, 2019). The value of nations is proved also by the immense numbers of immigrants seeking to leave corrupt or failed states for nations with well-functioning governments committed to the well-being of their citizens.

57 Consider as an example current Supreme Court interpretations of the 2nd amendment to the U.S. Constitution, which prioritize personal gun ownership over public safety.

responsibilities to fulfill their purpose. IPAT highlights that rights are claims on scarce resources, an aspect that can no longer be ignored in a world in rapid ecological decline. To create societies where human rights are secure, all members have opportunities to flourish, and other species can continue to prosper, we will need much smaller human populations. It is past time to acknowledge this reality and craft population policies accordingly.

6 Conclusion

Overpopulation threatens massive suffering for billions of people and extinction for millions of species. It imperils life's flourishing, the ultimate value. These facts justify efforts to reduce human numbers as quickly as humanely possible, as a matter of justice between current and future generations, and between humans and other species. Such efforts should start not someday, somewhere else, but here and now, in your society and mine. Addressing population is only part of creating just and sustainable societies, but it is a necessary part. "I" really does equal "P × A × T." While taking up population may prove contentious and challenging, continuing to ignore it will likely prove much worse.

Acknowledgements

The author thanks Michel Bourban, Kristin Cafaro, and the editors of this volume for detailed comments and suggestions which greatly improved the style and substance of this article, and Pernilla Hansson for creating and updating Figures 12.3–12.5.

References

Attenborough, D. (2011) "Impact of population growth on the planet," lecture to the Royal Society for the Encouragement of Arts, Manufactures and Commerce, London.

Beck, R. and Kolankiewicz, L. (2000) "The environmental movement's retreat from advocating U.S. population stabilization (1970–1998): a first draft of history," *Journal of Policy History* 12: 123–56.

Bradshaw, B. et al. (2021) "Underestimating the challenges of avoiding a ghastly future," *Frontiers in Conservation Science* 1: 615419.

Cafaro, P. (2013) "Reducing consumption to avert catastrophic global climate change: the case of aviation," *Natural Science* 5: 99–105.

Cafaro, P. (2014) "Avoiding catastrophic climate change: why technological innovation is necessary but not sufficient," in R. Sandler (ed.), *Ethics and Emerging Technologies*. Palgrave-Macmillan, 424–38.

Cafaro, P. (2015) *How Many Is Too Many? The Progressive Argument for Reducing Immigration into the United States*. University of Chicago Press.

Cafaro, P. (2021) "Just population policies for an overpopulated world," *Ecological Citizen* 5: epub-046–1.

Cafaro, P. (2022) "Reducing human numbers and the size of our economies is necessary to avoid a mass extinction and share Earth justly with other species," *Philosophia* 50: doi 10.1007/s11406-022-00497-w.

Cafaro, P. and O'Sullivan, J. (2019) "How should ecological citizens think about immigration?" *Ecological Citizen* 3: 85–92.

Cafaro, P. et al. (2022) "Overpopulation is a major cause of biodiversity loss and smaller human populations are necessary to preserve what is left," *Biological Conservation* 272: 109646.

Carens, J. (2013) *The Ethics of Immigration*. Oxford University Press.

Ceballos, G. et al. (2015) "Accelerated modern human – induced species losses: entering the sixth mass extinction," *Science Advances* 1: e1400253.

Cripps, E. (2015) "Climate change, population, and justice: hard choices to avoid tragic choices," *Global Justice: Theory Practice Rhetoric* 8: 1–22.

Cripps, E. (2021) "Population ethics for an imperfect world: basic justice, reasonable disagreement, and unavoidable value judgements," *Journal of Development Studies* 57: 1474.

Crist, E. (2019) "Decoupling the global population problem from immigration issues," *The Ecological Citizen* 2: 149–50.

Crist, E. et al. (2017) "The interaction of human population, food production, and biodiversity protection," *Science* 356: 260–4.

Crist, E. et al. (2021) "Protecting half the planet and transforming human systems are complementary goals," *Frontiers of Conservation Science* 2: 761292.

Daly, H. and Farley, J. (2010), *Ecological Economics: Principles and Applications*, 2nd edition. Island Press.

Dasgupta, P. (2019) *Time and the Generations: Population Ethics for a Diminishing Planet*. Columbia University Press.

Diaz, S. et al., (2019) "Pervasive human-driven decline of life on earth points to the need for transformative change," *Science* 366: 1327.

Dinerstein, E. et al. (2017) "An ecoregion-based approach to protecting half the terrestrial realm," *Bioscience* 67: 534–45.

Donaldson, S. and Kymlicka, W. (2011) *Zoopolis: A Political Theory of Animal Rights*. Oxford University Press.

Ehrlich, P. and Holdren, J. (1972) "One-dimensional ecology," *Bulletin of the Atomic Scientists*: 16–27.

Ganivet, E. (2019) "Growth in human population and consumption both need to be addressed to reach an ecologically sustainable future," *Environment, Development and Sustainability* 22: 4979–98.

Gheaus, A. (2016) "The right to parent and duties concerning future generations," *Journal of Political Philosophy* 24: 487–508.

Götmark, F. et al., (2021) "Discussing population concepts: overpopulation is a necessary word and an inconvenient truth," *Indian Journal of Population and Development* 1: 51–60.

Hardee, K. et al. (2013) *Voluntary Family Planning Programs That Respect, Protect, and Fulfill Human Rights: A Conceptual Framework*. Futures Group.

Hayward, T. (2016). "Ecological space: the concept and its ethical significance," in S. Gardiner and A. Thompson (eds.), *The Oxford Handbook of Environmental Ethics*. Oxford Press, 311–20.

Hedberg, H. (2020) *The Environmental Impact of Overpopulation: The Ethics of Procreation*. Routledge.

Hendrixson, A. et al. (2019) "Confronting populationism: Feminist challenges to population control in an era of climate change," *Gender, Place & Culture* 27: 307–15.

Hickey, C. et al. (2016) "Population engineering and the fight against climate change," *Social Theory and Practice* 42: 845–70.

Higgs, K. (2017) "Limits to growth: human economy and planetary boundaries," *Journal of Population and Sustainability* 2: 15–36.

Intergovernmental Panel on Biodiversity and Ecosystem Services (IPBES) (2019) *Summary for Policymakers: Global Assessment Report on Biodiversity and Ecosystem Services*. IPBES Secretariat.

Intergovernmental Panel on Climate Change (IPCC) (2022) *Climate Change 2022: Mitigation of Climate Change. Contribution of Working Group III to the Sixth Assessment Report of the Intergovernmental Panel on Climate Change*. IPCC Secretariat.

International Union for the Conservation of Nature (IUCN) (2020) "Importance for the conservation of nature of removing barriers to rights-based voluntary family planning," motion passed at IUCN World Conservation Congress.

Kant, I. (1785) *Groundwork of the Metaphysics of Morals*. Numerous editions.

Kolankiewicz, L. et al. (2022) *From Sea to Sprawling Sea: Quantifying the Loss of Open Space in America*. Numbers USA.

Kuhlemann, K. (2018) "'Any size population will do?': The fallacy of aiming for stabilization of human numbers," *The Ecological Citizen* 1: 181–9.

Leopold, A. (1966) *A Sand County Almanac with Essays on Conservation from Round River*. Oxford University Press.

Lewis, S. (2022) "Scientists have just told us how to solve the climate crisis – will the world listen?" *The Guardian*, April 6, 2022.

Lianos, T. and Pseiridis, A. (2016) "Sustainable welfare and optimum population size," *Environmental Development and Sustainability* 18: 1679–99.

Locke, H. (2015) "Nature needs (at least) half: A necessary new agenda for protected areas," in G. Wuerthner et al. (eds.), *Protecting the Wild: Parks and Wilderness, the Foundation for Conservation*. Island Press.

Meijers, T. (2016) "Climate change and the right to one child," in G. Bos and M. Düwell (eds.), *Human Rights and Sustainability*. Routledge.

Mill, J.S. (1859) *On Liberty*. Numerous editions.

O'Neill, B. et al. (2012) "Demographic change and carbon dioxide emissions," *The Lancet* 380: 157–64.

Overpopulation Project (2018) "Population growth is a threat to the world's climate," Gothenburg, Sweden.

Overpopulation Project (2020) "Rewilding success stories," Gothenburg, Sweden.

Pereira, H. and Navarro, L. (eds.) (2015), *Rewilding European Landscapes*. Springer International.

Population Institute (2015) *Demographic Vulnerability: Where Population Growth Poses the Greatest Challenges*. Washington, D.C.

Rees, W. (2020) "Ecological economics for humanity's plague phase," *Ecological Economics* 169: 106519.

Rieder, T. (2016) *Toward a Small Family Ethic: How Overpopulation and Climate Change Are Affecting the Morality of Procreation*. Springer.

Richardson, J. et.al. (2023). "Earth beyond six of nine planetary boundaries." *Science Advances* 9: 37.

Ripple, W. et al. (2020) "World scientists' warning of a climate emergency," *Bioscience* 70: 8–12.

Ripple, W. et al. (2021) "World scientists' warning of a climate emergency 2021," *Bioscience* 71: 894–8.

Rolston, H. III (1994) *Conserving Natural Value*. Columbia University Press.

Shragg, K. (2015) *Move Upstream: A Call to Solve Overpopulation*. Freethought House.

Staples, W. III and Cafaro, P. (2012) "For a species right to exist," in P. Cafaro and E. Crist (eds.), *Life on the Brink: Environmentalists Confront Overpopulation*. University of Georgia Press.

Tamir, Y. (2019) *Why Nationalism*. Princeton University Press.

Tucker, C. (2019) *A Planet of 3 Billion: Mapping Humanity's Long History of Ecological Destruction and Finding Our Way to a Resilient Future*. Observatory Press.

United Nations (2022) *World Population Prospects 2022*. Population Division, Department of Economic and Social Affairs, New York.

Van Vuuren, D. et al. (2018) "Alternative pathways to the 1.5 C target reduce the need for negative emission technologies," *Nature Climate Change* 8: 391–7.

Walsh, B. et al. (2017) "Pathways for balancing CO_2 emissions and sinks," *Nature Communications* 8: 14856.

Wiedmann, T. et al. (2020) "Scientists' warning on affluence," *Nature Communications* 11: 3107.

World Wildlife Fund (2022) *Living Planet Report 2022—Building a Nature-Positive Society*. Gland, Switzerland.

Wynes, S. and Nicholas, K. (2017) "The climate mitigation gap: education and government recommendations miss the most effective individual actions," *Environmental Research Letters* 12: 074024.

CHAPTER 13

Conclusion: Sufficiency-Driven Change

Tina Nyfors and Toni Ruuska

Human-induced impacts on the environment keep on escalating. The proposed geology of humankind, the Anthropocene, is characterized by the Great Acceleration with devastating environmental consequences signifying a worsening ecological overshoot and planetary ecological crisis. As argued in this the edited volume, the overshoot and ecological crisis result ultimately from growth, in material affluence, in deployment of resource-intensive technology, and in human population. The most severe repercussions of the deepening ecological crisis include climate change with rising global temperatures, superstorms, draughts, and heatwaves, and biodiversity loss signifying the sixth mass extinction event, which may render up to one million species extinct within decades. The ecological catastrophe is paired with humanitarian and geopolitical crises, as deep socio-economic inequality persists globally. Without decisive interventions addressing the root causes of our planetary crises, there is little hope for change for the better.

This edited volume has been written precisely because of this troubling situation. The conditions for sustainability and social justice have been deeply violated with perpetual growth in affluence (A), technology (T), and population (P). As a counter motive and ethos to growth and overshoot we have proposed sufficiency. However, and as we have argued in the Introduction, we have not treated the three factors (A, T & P) as equals, as we proposed the reorganization of the I = PAT equation to the I = ATP form. In the world where only a tenth of the humans cause half of all CO_2 emissions, the importance of social justice for sufficiency and sustainability needs to be emphasized. For instance, per capita income level and material standard of living, as well as the level of technologization, contribute directly to CO_2 emissions, and ecological impact more broadly. Hence, we stress the need to address the questions concerning too much affluence and inequality, and too much technology and inequality, before discussing the questions related to population. At the same time, we do not wish to sideline the population question, because it certainly is a sustainability concern as highlighted by Part 4 of this book. But given the prevailing socio-economic inequalities, it makes little sense to address population separately from material affluence and technological deployment.

 | DOI:10.1163/9789004746404_014

The chapters of this edited volume are primarily concerned with a context of growth and overshoot, where sufficiency implies decrease, moderation, being well off—or better off—with less, and the notion of enoughness as opposed to excess. As the editors of this book, we would like to further emphasize that the implications of sufficiency depend on one's starting point and circumstance. In the case of addressing perpetual economic growth, overshoot, and environmental problems, there is no doubt that degrowth is the primary concern today—another key motivation behind this book. In addition to respecting biophysical limits, sufficiency, as a notion and ethos of enoughness, relate to avoiding deprivation, that is, appreciating social limits and considering questions of social justice. Hence, enoughness can be understood to include both an ecological maximum and a social minimum, and then sufficiency is the space between them.

Sufficiency, can thus be, for instance, linked to the consumption corridor approach, which surfaces three starting points: a person can be situated above the corridor, within the corridor, or below the corridor. Being situated above the corridor signifies (different degrees of) overconsumption. For the richest tenth of the human population, which causes disproportionate environmental harm around the globe, sufficiency implies restraint and degrowth. In contrast to these overshooters, or, the over-consuming class, the people within the corridor constitute the sufficiency class. It is worth keeping in mind that this class represents more than a third of the human population with basic needs met while causing only moderate environmental impacts. In terms of bringing about sufficiency-driven change, the relevance of this class lays, for instance, in offering alternative visions and practices of a good life, compared to the carbon- and resource-intensive ways of living of the overconsuming class. Additionally, to the overshooters and the sufficiency class, one third of the human population is struggling to fulfil their basic needs. Here, the implication for sufficiency is, of course, to have their needs met in such a way that these people make their way to the sufficiency class.

In parallel to recognizing the underlying inequality and disparities of the global economy, we want to stress the importance of sufficiency as an ethos. Here we are in great debt to the work of Thomas Princen. According to Princen (see Chapter 2), sufficiency should be a guiding principle of sustainability transitions, especially in the face of declining resources. This is because the ethos of sufficiency challenges the dominant principles of growth, efficiency, and consumer sovereignty that are prevalent in modern industrial societies, by advocating for relational and ecological correspondence. Princen also calls for speaking differently, using sufficiency as part of a new language that counters the current norms of growth, expansion, and exploitation. Sufficiency is thus about both ecological and social limits and thresholds, but also an ethos of enoughness to guide societal change towards sustainability.

In this edited volume, the authors have also covered the dynamics of the current world economy, emphasizing how material affluence and technological deployment, particularly in the Global North, are often achieved at the expense of less privileged peoples and ecosystems around the world. This phenomenon is largely driven by the hegemony of capitalism as a system, which prioritizes capital accumulation, profit maximization, and technological advancement, and disregards questions of social justice and ecological sustainability. In Chapter 5 it was argued that capitalism and technological development contribute to a zero-sum world where the affluence and technologization of some is built on the exploitation of others labor and the degradation of natural habitats. This is evidenced by the scholarship surrounding the concept of ecologically unequal exchange, which demonstrates how wealthier nations, transnational corporations and other international organizations, and individuals, benefit from the disproportionate appropriation of resources from peripheries and less affluent regions, primarily located in the Global South.

Finally, this edited volume has called for a transition from the current growth-oriented and technological development -laden capitalist economic model to one that embraces degrowth, sufficiency and strong sustainability. Whereas sufficiency stresses enoughness, moderation, and balance, degrowth complements sufficiency by advocating for reductions in matter-energy throughput for the overly affluent, and the reorganization of societies to do well without perpetual economic growth. Sufficiency and degrowth thus both point to a shift towards a good life within limits, by highlighting the importance and meaningfulness of non-material values and equitable resource distribution. From the perspectives of sufficiency and degrowth, affluence and technology need to be re-thought, not as personal wealth and status symbols, or war machinery, like in capitalism, but as communal and inter-generational means for a good life in the post-capitalist futures. In the meanwhile, the questions concerning sufficiency in population need to be coupled with these deliberations in a way that secures enough for fulfilling human lives and leaves enough room and time for non-human flourishing.

The contributors to this volume do not necessarily share similar point of views or approaches to the great problems of our time, which should also inform us that there are not easy fixes available. Still, the contributors are brought together by their visions of enoughness that are linked with the A, T, and P factors. This means too that we have taken the questions of affluence, technology, and population as root causes to the current ecological overshoot, and as key obstacles for sustainability. In the light of this, we have collected below the key arguments from each of the contributors.

Part 1 (Setting the Stage) provided us with an overview of the idea of enoughness and relational sufficiency. Princen (Chapter 2) argued that sufficiency, or the idea of limits and enoughness, is an imperative for our time. Excess—or "too-muchness"—was properly showcased as the problem of modern societies, and sufficiency was presented as one, perhaps the best, alternative to excess. Jalas (Chapter 3) advanced the idea of IPACT equation, in which a consumption factor (C) is added to the analysis. This brings in a temporal component that points toward slower lifestyles, thus assisting humanity to tread more lightly upon the Earth.

Part 2 (Sufficiency in Affluence) considered the topics of growth and overshoot, capitalism, and degrowth, while working on the questions of sufficiency in affluence. Nyfors and Linnanen (Chapter 4) studied critically the numbers of overshoot by forcefully demonstrating that the wealthiest classes of humanity produce the largest footprints and thus have the greatest responsibility of sufficiency-driven change. Ruuska (Chapter 5) argued that the modern capitalist economy is based on growth, competition, and capital accumulation, ultimately pointing to the exploitation of humans and non-human nature. He called for degrowth transitions and practices towards sufficiency, that is, actions reducing the matter-energy throughput and its consequent impacts on the environment in a fair and planned manner. Nesterova and Jungell-Michelsson (Chapter 6) also supported degrowth actions while placing greater emphasis on business and deep transformation that can move us toward sustainable societies.

Part 3 (Sufficiency in Technology) explored technology and its impacts critically. Vadén (Chapter 7) highlighted how dependent yet blind we are to oil and other fossil fuels. For him, the concept of sufficiency is vital for us to break out of our "nafthist fog" and relearn how to live with the flows and rhythms of nature. Bonnedahl (Chapter 8) called for a shift from a wants-based or demand-oriented society to one that satisfies basic needs within a context of ecological sufficiency. He demonstrated how high-tech options are most often luxuries that do not fit any reasonable definition of a human need. Mohorčich (Chapter 9) studied human transport, energy demand, and consequently the need to move to a system of slow travel. All current high-speed travel options (such as trains and motor vehicles) are energy-intensive and thus highly wasteful and not sustainable. He thus called for a reorientation of society to slower and less energy-consuming modes of transport, such as bicycles, electric scooters, and walking.

Finally, Part 4 (Sufficiency in Population) discussed critically but constructively the difficult questions linked to population from the perspective of sufficiency and sustainability. Bourban (Chapter 10) argued that while the affluence and technology factors are important, population has been systematically overlooked, and that the question of population reduction must be seriously addressed if we are to reduce overall environmental impact. He thus advocated for a program of population degrowth. Heikkurinen and Uotinen (Chapter 11) commended the social shift to embrace environmental sustainability, but at the same time highlighted the neglect of the A, T, and P as root causes of the ecological crisis. Using Finland as a case study, they demonstrated the importance of raising the subject of overpopulation in national dialogues by offering a checklist to a more proper discourse on population. Finally, Cafaro (Chapter 12) examined critically questions of why it is so difficult to even talk about overpopulation. He concluded that to move towards sufficiency in population we must, for instance, improve family planning options, and discourage people from having more than one child.

In this book the authors have presented sufficiency as a response to the ecological overshoot and socio-economic inequality driven by growth in affluence, technology, and population. In contrast to growth and overshoot, sufficiency represents a balanced approach to ecological sustainability and social justice through which we may seek to meet human needs within planetary boundaries. Sufficiency also challenges the logic of efficiency, which often supports economic growth, by suggesting absolute reductions in consumption and production. Sufficiency is thus both a means and an end.

To find pathways towards sufficiency and sustainability, the authors have stressed the importance of addressing affluence, technology, and population in nuanced ways by acknowledging historical exploitation and disparities. By calling for degrowth and redistribution in affluence, rethinking technology beyond the silver-bullet metaphor, and approaching population discussions sensitively yet critically, this volume has tried to tackle some of the greatest challenges of our era. As the authors see it, sufficiency provides an alternative ethos that goes against the imperatives of growth-driven economics and growth-dependent societies. Surely, sufficiency might be perceived as an ill fit in the current political climate and deemed as impossible to implement because it entails reductions in a time when growth is still hegemonic. Yet, the authors of this volume find that challenging the growth-driven status quo is exactly what is needed now, when transformative change is desperately needed. On this endeavor, the humble notion of enoughness can serve as a powerful guiding star and drive societal change forward in an ecologically sustainable and socially just manner.

Index

www.ingramcontent.com/pod-product-compliance
Lightning Source LLC
Chambersburg PA
CBHW071248230926
51849CB00001B/265
* 9 7 8 9 0 0 4 7 4 6 3 9 8 *